Contents

Preface

The text of *Ground Studies for Pilots* has been completely revised and brought up to date. It is now produced in four volumes:

Volume 1 Radio Aids
Volume 2 Plotting and Flight Planning
Volume 3 Navigation General, Instruments and Compasses
Volume 4 Meteorology

The text, examples and exercises have been completely revised in line with the current syllabuses for the professional pilots' licences (Commercial Pilot and Airline Transport Pilot) and with aviation practice of the 1990s. However, as stressed by the original authors (sadly, both now deceased), accuracy combined with speed is essential in the subjects covered by this particular volume and this is still true, both operationally and under examination conditions.

I would like to acknowledge particularly the valuable assistance of my former colleague Tony Palmer in writing this volume. The portions of AERAD charts are reproduced by kind permission of British Airways. I am grateful too, for permission from Jeppesen & Co. GmbH to reproduce their material in Appendix 3 and for the Civil Aviation Authority's permission to reproduce sections of the CAA Instructional Plotting Chart–Europe for the diagrams in Chapter 2. Copies of the charts may be purchased from the addresses given in Appendix 3.

Examination practice questions are available by post from the Civil Aviation Authority, Printing & Publications Services, Greville House, 37 Gratton Road, Cheltenham, Glos GL50 2BN.

Hamble

Roy Underdown

GROUND STUDIES FOR PILOTS

Fourth Edition
Volume 2

PLOTTING AND FLIGHT PLANNING

S. E. T. Taylor
formerly British Airways and Chief Ground Instructor,
Malaysia Air Training, Kuala Lumpur;
Chief Ground Instructor, London School of Flying

and

H. A. Parmar
formerly Senior Tutor, Bristow Helicopters Ltd,
Chief Ground Instructor, Malaysia Air Training, Kuala Lumpur and
Specialist Instructor, London School of Flying

Revised by
R. B. Underdown
FRMetS MRAeS
formerly Director of Ground Training and latterly Principal,
College of Air Training, Hamble

BSP PROFESSIONAL BOOKS
OXFORD LONDON EDINBURGH
BOSTON MELBOURNE

First published in Great Britain by Crosby Lockwood & Son Ltd 1970
Reprinted 1972
Second edition 1974, reprinted 1976, 1978
Third edition, in three volumes, 1979
Reprinted by Granada Publishing 1981, 1983
Fourth edition published by Collins Professional and Technical Books 1986
Reprinted with updates by BSP Professional Books 1989, 1990

British Library
Cataloguing in Publication Data

Taylor, S. E. T.
Ground studies for pilots. – 4th ed.
Vol. 2: Plotting and flight planning
1. Commercial aircraft. Flying – Manuals
I. Title II. Parmer, H. A.
III. Underdown, R. B.
629.132'5216

ISBN 0–632–02481–X

BSP Professional Books
A division of Blackwell Scientific Publications Ltd
Editorial Offices:
Osney Mead, Oxford OX2 0EL
(Orders: Tel. 0865 240201)
25 John Street, London WC1N 2BL
23 Ainslie Place, Edinburgh EH3 6AJ
3 Cambridge Center, Suite 208, Cambridge MA 02142, USA
107 Barry Street, Carlton, Victoria 3053, Australia

Printed and bound in Great Britain by the Alden Press, Oxford

Section 1

THE EARTH

1: Form of the Earth

The Earth

The Earth is not a true sphere but is flattened slightly at the poles. The more correct description of its shape may be an ellipsoid. Its Equatorial diameter of 6 884 nm exceeds its Polar diameter by about 23 nm. This flattening is known as Compression, which is merely the ratio of the difference between the two diameters to the larger diameter. Expressed in mathematical terms:

$$\text{Compression} = \frac{\text{equatorial diameter} - \text{polar diameter}}{\text{equatorial diameter}}$$

and its value approximates $\frac{1}{300}$. However, for our purposes, the Earth is a sphere.

Great Circle (G/C)

We all agree that a line which directly joins any two places on the Earth represents the shortest distance between them. Now, if we continue one end of this line in the same direction right round the Earth until it finally joins up at the other end, we find that the circle we have drawn just divides the Earth into two equal halves. Try it on an orange, keeping the knife blade at 90° to the skin. Putting the story in reverse we can state that the smaller arc of a great circle always represents the shortest distance between two places. This is all important from our point of view.

To define it, a Great Circle is a circle on the surface of the sphere whose centre is the centre of the Earth, whose radius is the radius of the Earth and which divides the Earth into two equal parts. Rather a lengthy one to learn, but know it and keep it in mind when dealing with Great Circle problems. The definition in fact tells us more about the nature of a great circle than just its contents:

(1) that only one Great Circle could be drawn through any two places – try again on an orange;

(b) but if those two places were diametrically opposite an infinite number of Great Circles could be drawn. Lines joining the two Poles are examples.

The Equator and all the lines of meridians (longitudes) are examples of Great Circles (although, technically, meridians are semi-great Circles).

Small Circle

A Small Circle stands in contrast to a Great Circle. By definition, any circle on the surface of the Earth whose centre and radius are not those of the sphere itself is a Small Circle. All parallels of latitudes (except the Equator) are Small Circles. They do not represent the shortest distance between two places.

Latitude and Longitude

A reference system in international use of which you have no doubt heard. First of all, a Great Circle is drawn round the Earth through the North and South Poles passing through Greenwich. That half of the Great Circle between the two Poles which passes through Greenwich is called the Prime or Greenwich Meridian. The other half is called the Greenwich anti-meridian. The Greenwich meridian is labelled 0° and its anti-meridian, 180°. Thus, with this E – W division established, more Great Circles in the form of meridians could be drawn, both to the east of Greenwich and to the west.

The next step is to have a datum point for N – S divisions. This is obtained by dividing the Earth by a Great Circle mid-way between the two Poles, all points on it being equidistant from the Poles. Such a Great Circle, to your surprise, is called the Equator, and labelled 0° Latitude. Small Circles are now drawn, parallel to the Equator, towards both poles – these are parallels of latitude.

Definition of Latitude: it is the arc of a meridian intercepted between the Equator and the reference point. It is measured in degrees, minutes and seconds, and is termed North or South according to whether the place is to the north or south of the Equator.

Definition of Longitude: longitude is the shorter arc of the Equator intercepted between the Greenwich meridian and the reference point. It is measured east or west of the Prime meridian in degrees, minutes and seconds.

It is the meridians themselves that indicate North – South direction: the parallels run East – West.

The whole network of Latitude and Longitude (also called parallels and meridians) imagined to cover the Earth is called a Graticule. Thus, on a complete graticule we would see meridians starting from Greenwich as 0° going right round to the East and West up to 179°59′59″E and 179°59′59″W. 180° is common. Similarly to N – S, we would have parallels right up to 90°N and S, the Poles. A degree is divided into 60 minutes, and each minute is divided into 60 seconds (1° = 60′; 1′ = 60″).

And while on the subject of latitudes and longitudes, there are two more definitions you ought to be familiar with. They are: Change of Longitude and Change of Latitude.

Change of Longitude: it is the smaller arc of the Equator intercepted between the meridians of the reference points. It is named East or West according to the direction of the change.

Change of Latitude: it is the arc of the meridian intercepted between the parallels of the two places and is named North or South according to the direction of the change.

In the following sketch, if the flight was made from A to B, the ch long (change of longitude) is 2°E; ch lat (change of latitude) is 5°N. If the flight was from B to A, the ch long is 2°W and ch lat 5°S.

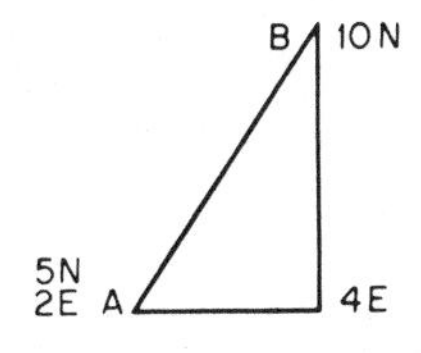

Fig. 1.1

Rhumb Line (R/L)

We established above that the shortest distance between any two places is along the Great Circle. This would be the ideal line (call it a 'Track') to fly. However, there is this disadvantage: the Great Circle from one point to another will cross the converging meridians at different angles. Since the meridians form the basis of our track angle measurements, this would mean continuous alterations to the track angles as the flight progresses.

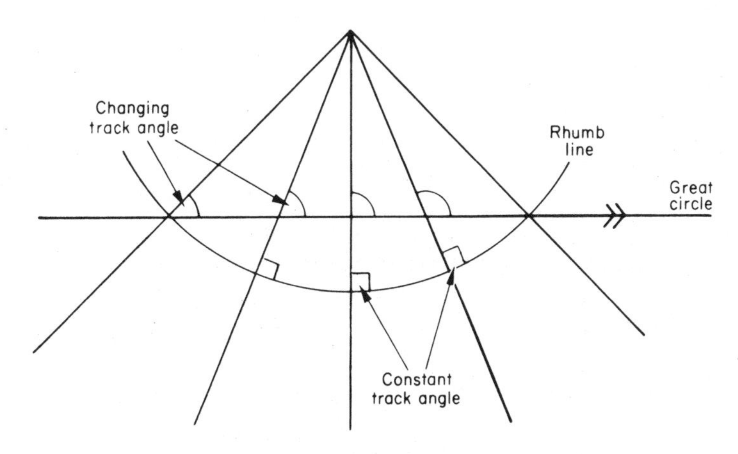

Fig. 1.2

Apart from a 090°/270° R/L (i.e. a parallel of latitude or the Equator), all others spiral towards the Poles. Their curvature and so their disparity from the equivalent G/C, which is the nearest approach possible to a straight line on the curved surface of the Earth, increases with latitude. It follows that, in low latitudes, G/C and R/L distances will show little difference. After all, the equator is both a G/C and a R/L. In high latitudes, however, there can be a great and wholly unacceptable difference.

R/L and G/C distances comparison

Consider a flight from 80°N 00° E/W to 80°N 180°E/W (Fig. 1.3). The G/C route is over the Pole and the distance will equal 20° of latitude or 20 x 60 = 1 200 nm. Simple plane geometry suggests that the R/L distance around the parallel will be 600 π = 1 885 nm. In fact, on the spherical earth as opposed to a flat diagram, the exact distance is 180 x 60 x cosine 80° = 1 875 nm.

R/L and G/C directions comparison

It is often said that flying a G/C, with a need to constantly change true direction

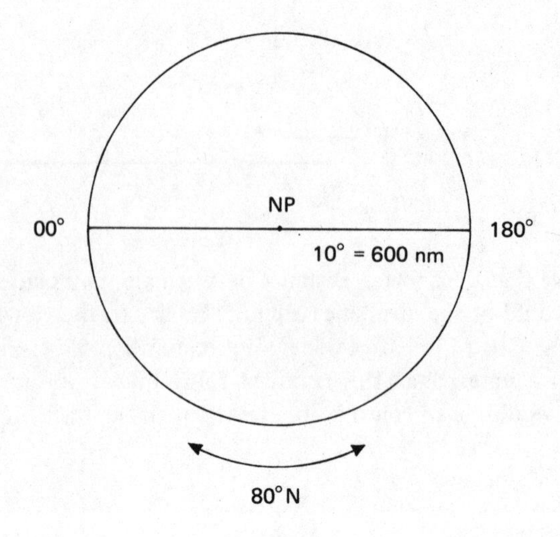

Fig. 1.3

unless flying due North or South (meridians are semi-Great Circles) or East or West along the Equator, creates practical steering problems. In practice, only aircraft with automatic (computerised) navigation systems are likely to fly true headings and these will quite easily direct the auto-pilot no matter how many changes are required.

Consider two North Atlantic routes:

Shannon – Gander (G/C 1715 nm R/L 1748 nm)

	Initial direction		Final direction		Direction change	
	(T)	(M)	(T)	(M)	(T)	(M)
G/C	281°	294°	245°	275°	36°	19°
R/L	263°	276°	263°	293°	0°	17°

Belfast – Keflavik (G/C 747 nm R/L 1748 nm)

	Initial direction		Final direction		Direction change	
	(T)	(M)	(T)	(M)	(T)	(M)
G/C	325°	336°	311°	337°	14°	1°
R/L	318°	329°	318°	344°	0°	15°

It will be seen that on some routes less change of magnetic track direction is involved when flying the G/C. Remember, also, the G/C always has the distance advantage. In any case, on long routes, regular alterations of heading will have to be made to conform with air traffic routes and/or to take account of varying wind conditions. These considerations further blur any clear distinction between the number of direction changes required to fly the two types of route.

Distances on the Earth's surface

The basic unit used for aviation track distances is the International Nautical Mile (nm) which is 1852 m in length. For practical purposes, it can be assumed to be the length of 1′ of G/C arc on a spherical Earth. The true earth is not a perfect sphere. Modern on-board automatic navigation computers make allowances in their calculations for the non-spherical shape of the Earth but the differences are not of any significance in any practical calculations that a pilot may have to make for himself.

The Equator is a G/C and so the 360° of longitude around it represents 360 x 60 = 21 600 nm. As meridians are semi-Great Circles, 1′ of geographical latitude can be assumed to be 1 nm. Do not fall into the trap of treating 1′ of longitude as a nm – this is only true at the equator but elsewhere it only equals cosine latitude nm (see volume 3). The distance along a meridian from the Equator to the Pole is 90 x 60 = 5400 nm.

The metric system provides for the decimalisation of the angular system with a right angle being divided into 100 grade (French for degree) and each grade being divided into 100 minuit (French for minute). Using this system, the minuit was established as the basic unit for large distances, i.e. the kilometre (km). It follows that the distance from the Equator to the Pole will be 100 x 100 minuit of latitude or 10 000 km and so:

10 000 km = 5 400 nm
and so 1 km = 0.54 nm
or 1.852 km = 1 nm
(1 852 m)

Conversions of distance units

The only other large distance unit encountered, and then only when dealing with the public, is the statute mile (sm). This is an arbitrary legal measure of 5 280 feet or approximately 1 610 m = 1.61 km.

1 nm = 1.852 km = 1.15 sm

These conversions are available on aviation circular slide rules and can easily be effected using the above values and the electronic calculator. Many maps give scale lines for all three units and these may be used as convenient conversion scales.

Most aviation work is done in nm and it should always be remembered that 1′ of geographical latitude for all practical purposes can be used as 1 nm and so, providing a map has a latitude scale that is reasonably well sub-divided, a simple nm scale line is always available.

For shorter distances (runway lengths, visibilities), the metre is gradually becoming the standard unit although feet are still commonly used for altitudes and elevations. The following conversions will be useful:

1 m = 3.28 feet or 39.37 inches
1 inch = 25.4 mm

Section 2
NAVIGATION PLOTTING

2: Basic Navigation Principles

These days it is very unusual to operate an aircraft using just the basic navigational techniques that will be described in this chapter. However, they represent the basic principles on which all navigation systems operate and should also be regarded as useful 'fall-back' procedures for use when the more sophisticated modern systems fail or become suspect.

The 'Art' of Navigation

All navigation is the 'art' of being able to give good answers to the following questions:

Where am I now?
How did I get here?
What am I going to do about it?

It is interesting to note that modern technology can give incredibly accurate answers to the first two questions but can only offer suggestions for the third. It is not given to a mere mortal to know what the future holds and so any method of forecasting it must make assumptions which may or may not prove to be valid. This element of uncertainty is why navigation is often referred to as an 'art' instead of a science. Certain basic procedures will have to be mastered before we can start practising this 'art'.

The Velocity Triangle

It is easy to appreciate that, to steer a motor-boat directly across a fast flowing river, it will be necessary to head the boat up-stream so that the combined velocities (speeds in defined directions) of the stream and the boat through the water will give a resultant velocity acting at 90° to the flow (Fig. 2.1).

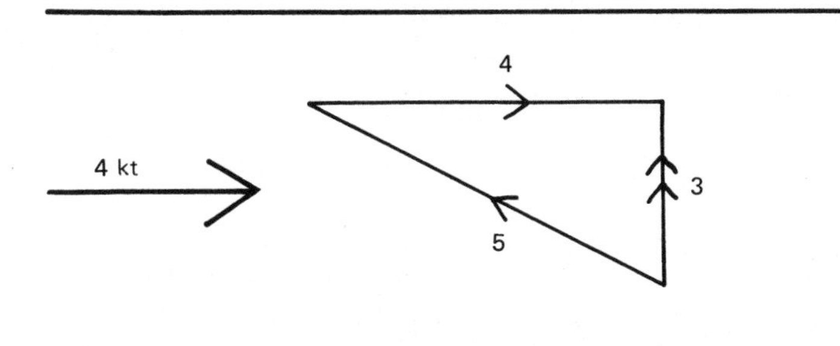

Fig. 2.1

The problem is solved by constructing a velocity triangle. Note that the two component velocities – the boat's 5 kt through the water and the water's flow of 4 kt form two sides of a triangle with the arrows indicating the directions of motion following each other round the triangle and opposing the direction indicated by the arrows on the resultant velocity of the boat's actual motion over the ground (i.e. the river bed). Note that the unit of speed we use is the knot which means one nautical mile per hour. This would seem to be appropriate when talking about boats but we will continue to use it for aviation because of the simple relationship of the nautical mile to Earth distances (see Chapter 1). The triangle could be solved by a scale drawing or by simple mathematics. In either case the answer would be obtained that the boat should be headed up-stream through an angle of 53° and the resultant speed across the river would be 3 kt. We could say that we are expecting the boat to be drifted 53° to starboard (right) and that its ground speed is 3 kt.

The Aircraft Velocity Triangle

The situation with an aircraft is exactly the same. The air mass in which it is flying will have motion in just the same way as the river. The motion of the air is given by the wind velocity (W/V) which is expressed by giving the direction from which the wind is blowing (W/D) and the speed of the air's motion over the ground (W/S) in knots (Fig. 2.2).

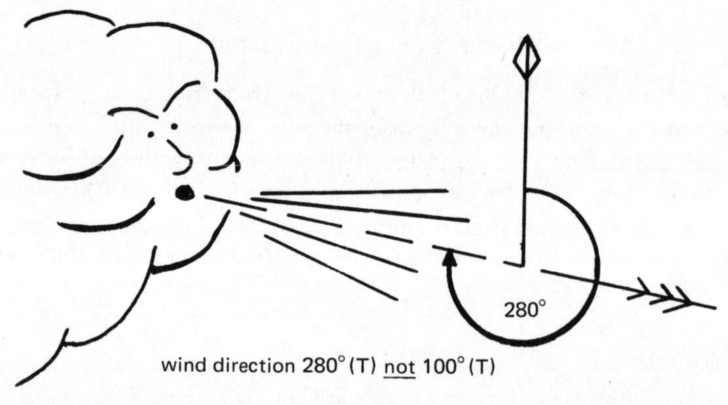

Fig. 2.2

A W/V of 280°/50 kt would indicate that a free balloon would move in a direction of 100° under its influence at a speed of 50 kt.

The equivalent of the boat's motion through the water is the aircraft's air velocity represented by the True Airspeed (TAS) and the True Heading (Hdg (T)) The aircraft's air velocity would equate to its motion over the ground only in calm or still air conditions.

True Airspeed (TAS)

In practice, this will usually be derived from the airspeed indicator (ASI) or the Machmeter – see volume 3 for details. Modern on-board automatic navigation systems will obtain TAS from the Air Data Computer (ADC). Aircraft with ADC may have TAS indicators but all aircraft will have the direct-reading, pressure-actuated ASI, as they play a very important part in pilotage techniques. Critical aircraft speeds are always expressed as Indicated Airspeeds (IAS), i.e. the speeds given by the ASI. Aircraft capable of high speeds will also carry Machmeters which are also direct-reading, pressure-actuated instruments. These will give the aircraft's TAS as a proportion of the speed of sound through the air for the particular temperature conditions.

Standard navigational computers such as the Airtour CRP 5 carry conversion scales which are specially designed to compute the TAS. Refer to the instructional handbook for your computer and note that by setting the corrected outside air temperature (COAT), sometimes referred to as the ambient temperature, against the pressure altitude (altimeter reading with 1013 mb on the sub-scale) it becomes possible to read off the TAS on the main circular slide rule outer scale against the Rectified Airspeed (RAS) on the inner scale. The RAS or the Calibrated Airspeed (CAS), as the Americans call it, is the IAS corrected for various instrument and installation errors (see volume 3). It will only equal TAS when the air has the calibration density (pressure 1013 mb, temperature +15°C). The computer makes the necessary density correction to produce the TAS.

Check you are doing it correctly with this example:

PA 10 000 ft COAT −15°C RAS 150 kt Solution 171 kt

Be careful if the TAS when calculated exceeds 300 kt. In this case an additional correction is required using an additional sub-scale on the computer. Computers usually refer to this as 'comp. corr.' (compressibility correction). Try this example:

PA 30 000 ft COAT −45°C RAS 250 kt

On a first setting a TAS of 400 kt will be obtained but, after adjustment using the comp. corr. scale, a more correct TAS of 394 kt will be obtained.

If you have an electronic navigational computer that has a program for calculating TAS, it will give the correct answer regardless of the TAS. You can test the truth of this statement by solving the last example on an electronic calculator. Incidentally, the CAA do not permit the use of this type of calculator in their examinations.

Obtaining TAS from Mach No.

Modern aircraft often cruise at an indicated Mach No. Again the standard navigation computers provide a method of obtaining the TAS. The COAT is set against a Mach Index on a special sub-scale and the TAS can then be read off on the main outer scale against the Mach No. on the inner scale of the circular slide rule. Try this example:

Mach No. 0.82 COAT −45°C TAS = 483 kt

An alternative solution, if you are using an electronic calculator, is to make use of this formula which is set out here for the calculator:

°C + 273 = √ x 39 x Mach No. = TAS

Remember, if required to make the temperature negative, to use the +/− key after putting in the temperature. Try the previous example by this method.

Now try this calculation by both methods in readiness for when you are flying Concorde:

COAT −57°C Mach No. 2.05 TAS 1 175 kt

Aircraft Heading (Hdg)

The direction in which the aircraft is pointing measured clockwise from a particular direction datum is known as the aircraft heading. In practice the following headings will be encountered:

Hdg(T)	True heading
Hdg(M)	Magnetic heading
Hdg(C)	Compass heading
Hdg(G)	Grid heading (see Chapter 5)

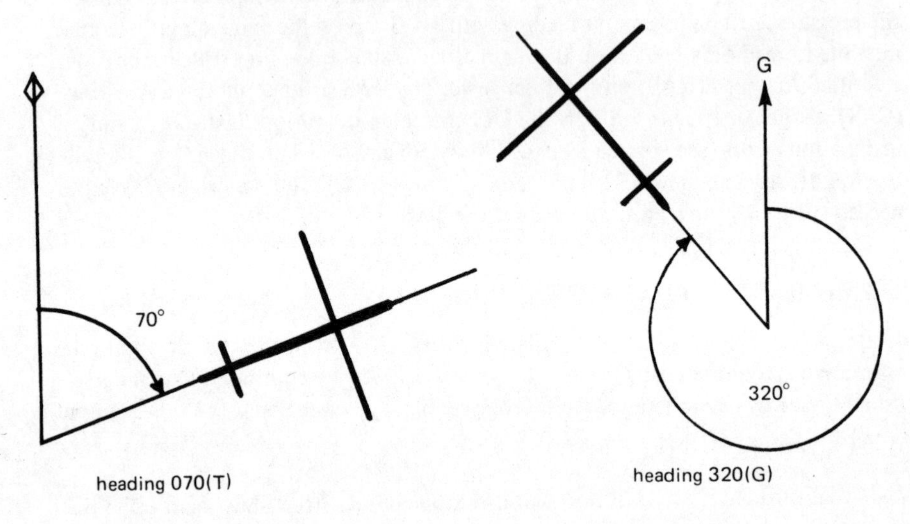

Fig. 2.3

To convert compass direction to magnetic direction, the correction referred to as deviation (see volume 3) should be applied and to convert magnetic direction to true direction, variation should be applied:

$$\text{(C)}\ \begin{matrix}+\text{ E devn}\\-\text{ W devn}\end{matrix}\ = \text{(M)}\ \begin{matrix}+\text{ E varn}\\-\text{ W varn}\end{matrix}\ = \text{(T)}$$

and so conversely:

$$\text{(T)}\ \begin{matrix}-\text{ E varn}\\+\text{ W varn}\end{matrix}\ = \text{(M)}\ \begin{matrix}-\text{ E devn}\\+\text{ W devn}\end{matrix}\ = \text{(C)}$$

See how this works with the examples illustrated in Figs. 2.4 and 2.5.

In modern compass systems, the deviations are usually very slight. Variation can be very large and will be obtained from the isogonals on the chart (see volume 3). It is important to check that the variation is up to date. A note in the margin of the chart or an annotation on the actual isogonals will indicate the year for which the variation is given. If more than two or three years out of date, it may be necessary to up-date the values using the corrections noted on the chart.

Track and Ground Speed (Tr and G/S)

In UK usage, track refers to the aircraft's actual movement over the ground. It may

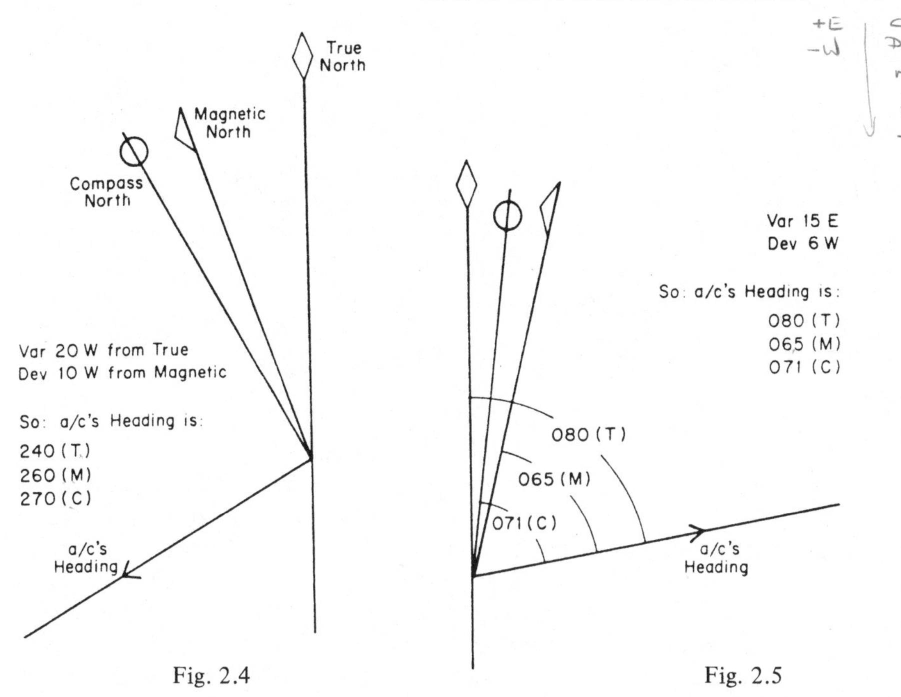

Fig. 2.4

Fig. 2.5

be the required or desired track, the calculated or DR (dead reckoning) track or the track already achieved usually referred to as the TMG (track made good).

In USA usage, and so often encountered when operating American aircraft and instrument systems, 'course' is used to indicate the required track and 'track' is reserved to describe the actual path over the ground. In radio navigation, a simple VOR set-up will have a CDI – a Course Deviation Indicator – which shows if the aircraft is deviating from the required track (course).

The G/S is the speed the aircraft is making good or is expected to make good (DR G/S) over the ground. Obviously in still air conditions (zero W/V), the TAS and the G/S will be the same. With a dead tail wind the G/S will be the sum of the TAS and W/S and with a dead head wind the G/S will be the difference of the TAS and the W/S.

It is a useful check on calculations to remember these two simple relationships for they place absolute limits on the G/S that may be found with given values of TAS and W/S and so can be used to reject impossible answers that may be presented.

The Aircraft Velocity Triangle

This is illustrated in Fig. 2.6. It will be noticed that the single arrow on the aircraft air velocity vector is followed round the triangle by the three arrows of the W/V vector. The two arrows on the aircraft ground velocity vector are in the opposite direction around the triangle. Try drawing this vector triangle for yourself using a simple and convenient scale such as 1 cm = 10 kt.

Hdg 090(T) TAS 180 kt W/V 040/45 kt

Solution Tr 103(T) G/S 156 kt

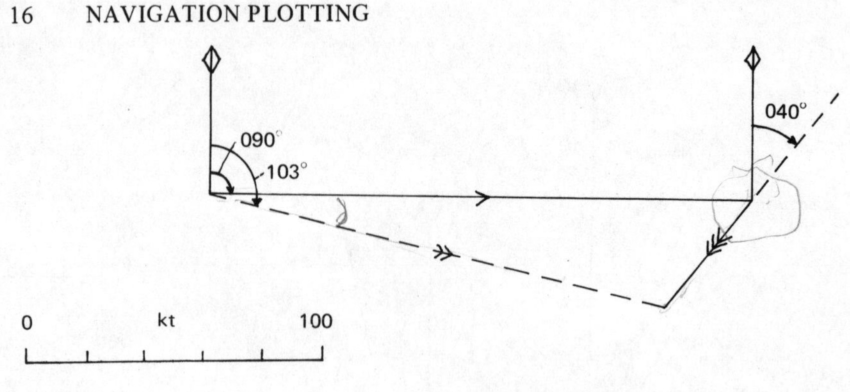

Fig. 2.6

Drift

Consider the significance of the vector triangle in Fig. 2.6. Try to think yourself into the pilot's seat – you would be pointing the aircraft's nose due East and the wind blowing from 040° (round about NE) would be coming at you from ahead and to your left. Naturally it would have the effect of 'drifting' you to the right and slowing you down. In this case the aircraft would be drifted 13° to the right (starboard) and the aircraft would move crabwise along a track of 103°. In Fig. 2.7, the movement of an aircraft with a heading of 256°(T) and a drift of 14°P giving a track of 242°(T) is illustrated.

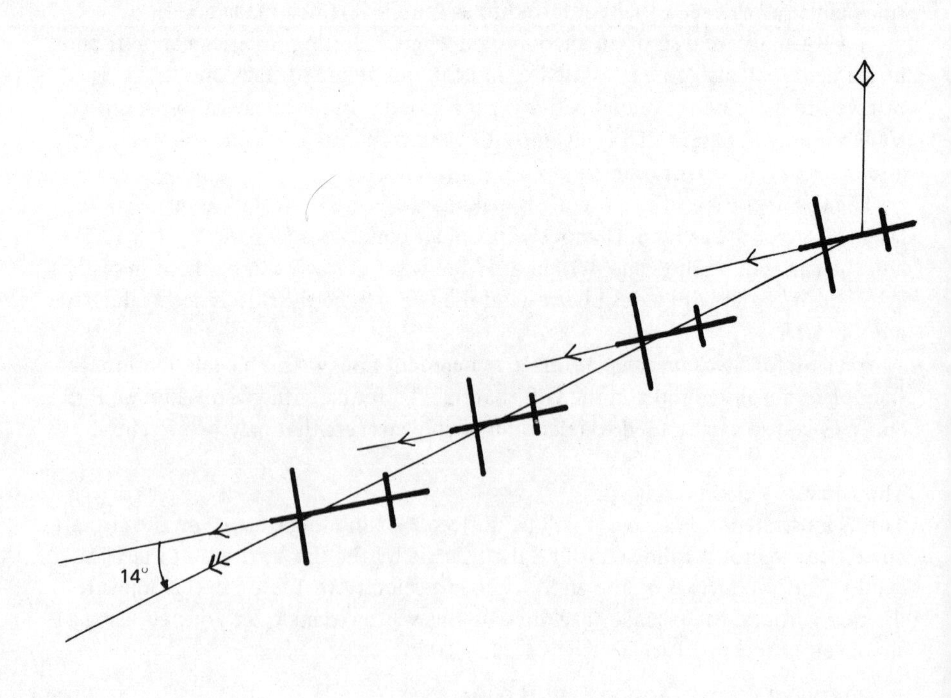

Fig. 2.7

Navigation Computers

In practice, the solution of the vector triangle is accomplished by using a specially designed computer. It may be of the analogue type which effectively provides a simple method of drawing the vector triangle – these are often referred to as Dalton computers after Dr Dalton who invented the system – and typical of these are the CRP series produced by Airtour International. A computer of this type is essential for the CAA exams. There are also electronic digital calculators available where it is only necessary to type in the basic information, e.g. TAS, W/D, W/S and Tr, and then ask the machine to compute the required heading and G/S. At present the use of these is not permitted by the CAA. By their nature they tend to be a little more accurate but they are certainly no more rapid than a Dalton in experienced hands. You must acquire a Dalton-type computer, study the handbook carefully and practise assiduously until it becomes second nature.

The heading and G/S problem

In practice, finding these is the commonest problem. Before flight, for example, a flight plan will require this problem to be solved for each stage of the flight.

To solve this problem by drawing requires a slightly different sequence of working. In Fig. 2.8, the track required is 225°(T), TAS 200 kt and W/V 280°/40 kt.

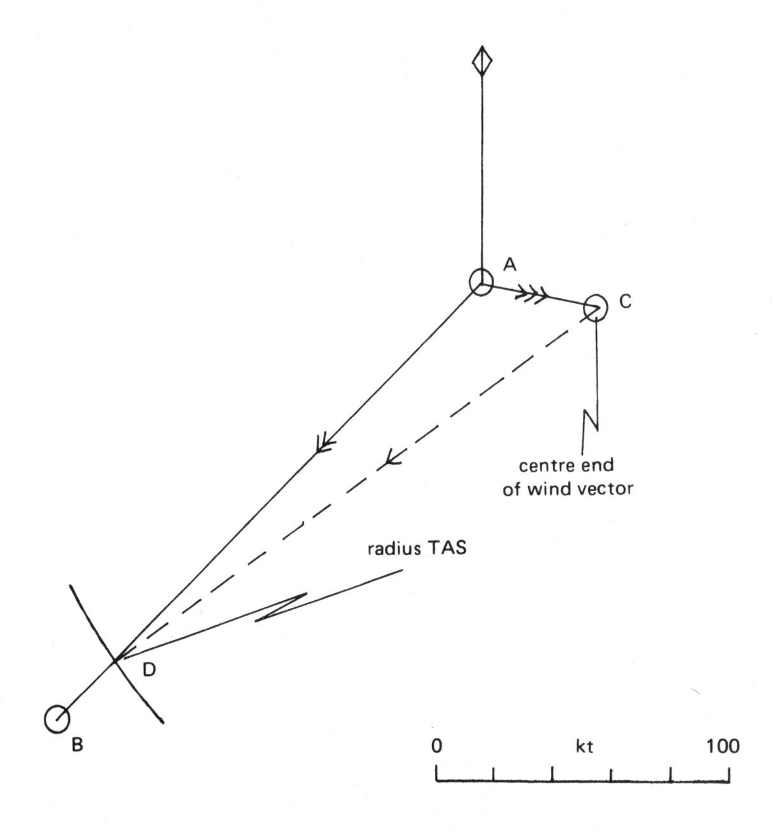

Fig. 2.8

Before we start on the solution, stop and think yourself into the situation. Sitting in the aircraft, the wind will be coming at you from a direction of 55° to your right (280 – 225) and so will cause left (port) drift and a G/S less than the TAS. Make sure that the final answer agrees with this.

The sequence of working is indicated by the letters. Try drawing it for yourself using a scale of 1 cm = 10 kt. The correct answers are: Hdg 234°(T) G/S 174 kt drift 9°P. Your answers should be within 2° and 4 kt. Now try the problem on your Dalton computer.

If AB represented on a chart the track required, a pilot would need to leave A and head his aircraft 9° to the right of the required track and rely on the wind to drift him 9° to the left so that the required track of 225°(T) is made good. Notice that as we foresaw the G/S is less than the TAS.

Wind Components (wc)

These are often referred to in aviation. Cross-wind components are usually identified as such although a more accurate description would be cross-track or cross-runway components. The other type of wind component is usually identified by the addition of 'head' or 'tail' or by the signs '–' or '+'. A word of warning here – in aircraft performance calculations when dealing with take-off and landing problems, the sign convention is reversed. The more common components – head – are given the sign '+' and the tail components are '–'.

Strictly, there are two distinct types of along-track or along-runway components:

'effective' wind components = G/S – TAS
'true' wind components = W/V resolved along the Tr or R/W direction (i.e. W/S x cosine wind angle)

In practice, the expressions 'effective' and 'true' are rarely seen. For flight planning purposes, the tables of wind components produced for a specified TAS are always 'effective' components. In performance manuals tables and diagrams give 'true' components.

(Effective) Wind Components

A few examples first. Ideally you should have a copy of the CAA Flight Planning Data Sheet 33 – this is essential for Chapter 8 – and also your Dalton computer. Start by calculating the G/S in these cases:

Tr	TAS	W/V	Answer (G/S)
050	380	140/120	360
050	480	140/120	466

In these cases the (effective) wind components would be:

360 – 380 = –20 kt or 20 kt head wind component (hwc)
466 – 480 = –14 kt or 14 kt hwc

Now compare these results with those given on pages 24 and 25 of Data Sheet 33 entering with a wind/track angle of 90° (140 – 050) and a wind speed of 120 kt. Note, that even in these conditions of maximum drift due to a high wind speed at 90° to track, the 'effective' component only changes by 6 kt with a change of TAS of 100 kt. It will be realised that at a TAS of 430 kt, the correct component would be −17 kt but if interpolation was ignored, the answer from either table in this extreme case would only be in error by 3 kt. In practice therefore, an effective wind table for a specified TAS can be used quite reasonably for any TAS within ±50 kt. Notice that a wind at 90° to track gives a significant effective head wind component and that, if the pilot flew on the reciprocal track, there would be exactly the same 'effective' head wind component. This demonstrates that an effective head wind component on a track does not necessarily indicate that there will be a corresponding tail wind component on the recriprocal track. Assuming this in the 380 kt TAS case would cause an error of 40 kt (400 − 360) when assessing the G/S on the reciprocal track.

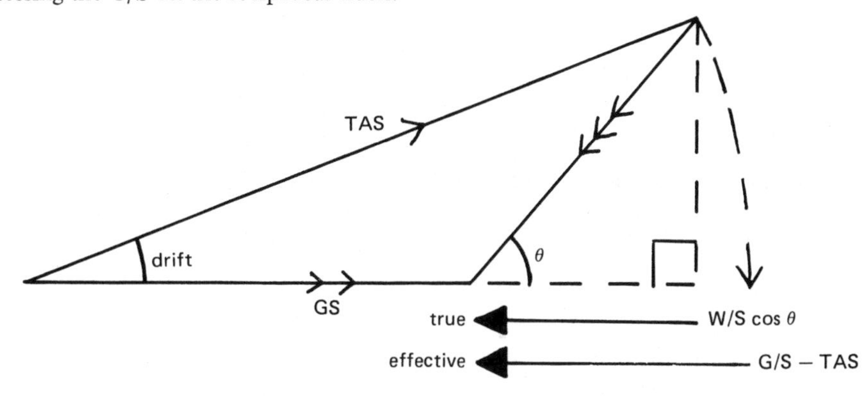

Fig. 2.9

(True) Wind Components

This is the component of the wind resolved along the track or, more commonly, the runway. Unlike (effective) wind components, (true) wind components are not dependent on the TAS. The other significant difference is that the (true) component is zero with a wind at 90° to the track whereas, the (effective) component can be quite a significant head wind as was seen in the previous paragraph. It will be noted in Fig. 2.9, which shows the difference between the two types of component, that the (effective) wind component is less favourable than the (true) wind component. Whenever there is any drift this will always be the case. Naturally, when flying up and down wind, there will be no drift and no difference between the two components, both of which will be equal to the wind speed. On the other hand, as already noted, the other extreme is with the wind at 90° to track when there will be no (true) component but there will be an (effective) head wind component.

(True) components are normally used for landing and take-off performance calculations where the cross-wind (90° to the runway) is also important. Both of these (true) components can be evaluated with the Dalton computer using the square graticule on the slide, or by scale diagrams or tables (both of which may be found in Flight Manuals), or by simple calculation on the electronic calculator:

along-runway component W/S x θ cosine =
across-runway component W/S x θ sine =

where θ is the difference between the wind direction and the runway.

The above formulae, which are written as they will be entered into the calculator, will give positive results in the case of head winds (θ less than 90° or more than 270°) but negative for tail winds (θ 090° through 180° to 270°). This accords with the standard practice for this type of calculation.

When evaluating runway components, it should be realised that runways are normally described by their magnetic directions and so the W/V should be converted into a magnetic direction also. This is demonstrated in the following example:

What are the along- and across-runway components of a W/V 090°T/40 kt on runway 07? Magnetic variation is 10°W.

Magnetic W/V 100°M/40 kt   Runway 070°M
$\theta = 100 - 070 = 30°$
Head wind = 40 cos 30° = 35 kt (nearest kt)
Cross wind = 40 sin 30° = 20 kt

Note that if landing in the opposite direction:

Headwind = 40 cos 150° (250 − 100) = −35 kt
indicating a 35 kt tail wind component.

Chart Work

In aviation in most latitudes, the commonest chart available for general use is the Lamberts Conformal Conic with two standard parallels. In equatorial regions, the standard Mercator chart may be encountered and in polar regions, the Polar Stereographic. Volume 3 gives greater detail about these charts. It can be assumed that, provided the use of the chart is confined to the areas for which it is best suited and that it is a conformal (orthomorphic) chart, the following properties may be assumed:

reasonably constant scale over one chart;
great circles practically straight lines;
angles are correctly represented (conformality).

Using the Lambert's Chart

For the purpose of this chapter, we will demonstrate navigation procedures on a Lambert chart – reference to using a Mercator or a Stereographic will be found in subsequent chapters. The chart used for demonstration purposes in this chapter will be the CAA Instructional Plotting Chart – Europe as used in their examinations. This is a 1:1 000 000 Lambert's and so a 20 inch (50 cm) 1:1 000 000 scale ruler reading in nm will prove very useful. In addition, a pair of dividers, a pencil compass, a 5 inch (13 cm) protractor and a navigational computer will be needed. Using an HB pencil will make it easier to rub out ready for re-use although fairly frequent use of a pencil sharpener may be required to keep a good working point. A draughtsman's clutch pencil will be found very satisfactory.

Plotting Positions on the Chart

The chart has a 30′ graticule, so the square protractor will be used to plot and to read off positions accurately. In Fig. 2.10 the corner of the protractor has been placed on the beacon ODN and carefully aligned with the chart graticule enabling the latitude 55°35′N and longitude 10°39′E to be read off. Conversely, the protractor could have been aligned to cut the graduations for 55°35′N and 10°39′E and the corner of the protractor would indicate the required position.

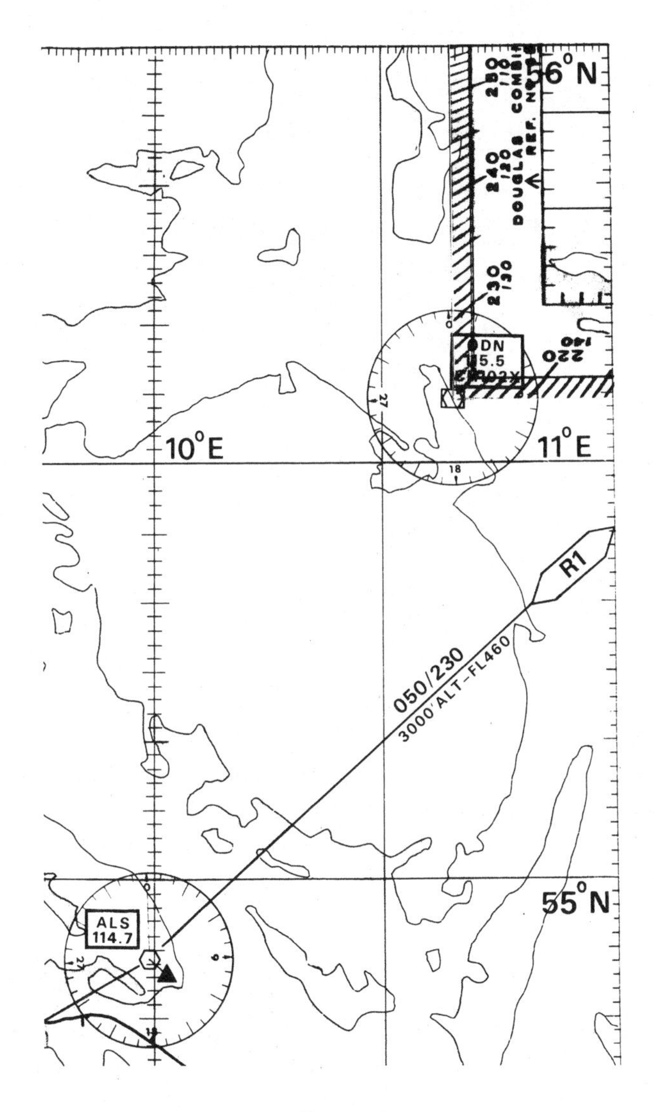

Fig. 2.10

Measuring distances and angles

In many cases on standard routes, directions and distances can simply be read off the chart. For example, in Fig. 2.11, the distance from VOR OSN to VOR RKN is 56 nm. Check this with your dividers using the latitude scale (1′ of latitude = 1 nm) and also using your scale ruler.

The magnetic track can also be read as 269°. Alternatively, measuring the track at the mid-meridian (7°30′E) to be 266°(T) and applying the local variation of 3°W will give the same result. Notice also that the approximate magnetic direction could be read from the compass rose around the OSN VOR which is orientated to Magnetic North.

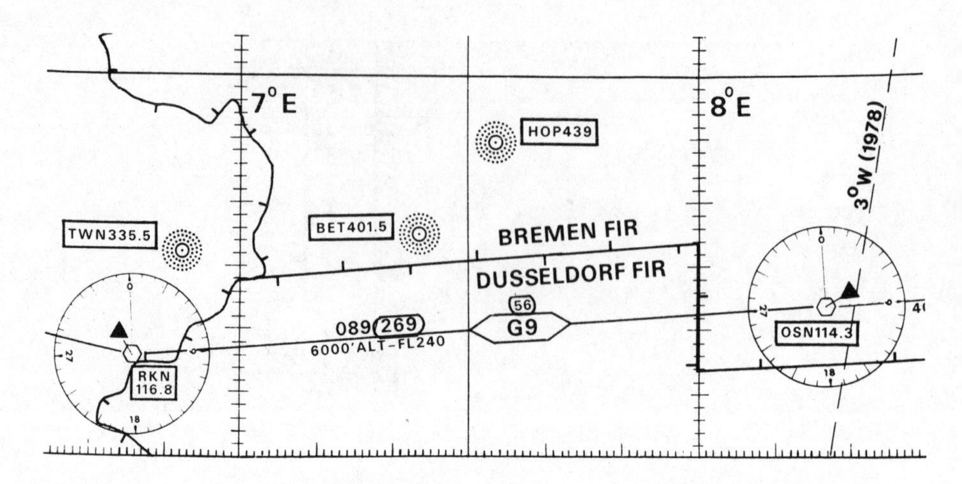

Fig. 2.11

To improve on this accuracy, try aligning the protractor with Magnetic North as indicated by the compass rose and then reading the track direction. Although this technique will not be used for measuring tracks that already have the answer printed on the chart, it will be used for plotting radio bearings.

Plotting Radio Bearings

As an essential part of fixing an aircraft's position, it could well be required to plot position lines derived from radio bearings or ranges. The information that will normally be available to any aircraft equipped for IFR flight will be:

VHF or VOR bearings (QDM or QDR)
ADF bearings (relative or magnetic)
Radar/DME ranges

(Refer to volume 1 for more details.)

All of these will give a position line – that is a line along which the aircraft is believed to lie at a given time. It is customary, having drawn this line on the chart, to put a single arrow at each end of it and write the time along it.

Range Position Lines

These will be derived from DME or by airborne radar measuring the range of an identifiable ground feature such as a small island or headland. Plotting is just a matter of drawing the arc of a circle with a radius of the range measured around the DME beacon or ground feature (see Fig. 2.12).

Plotting VHF or VOR bearings

Basically these are G/C bearings measured at a ground station – usually the station

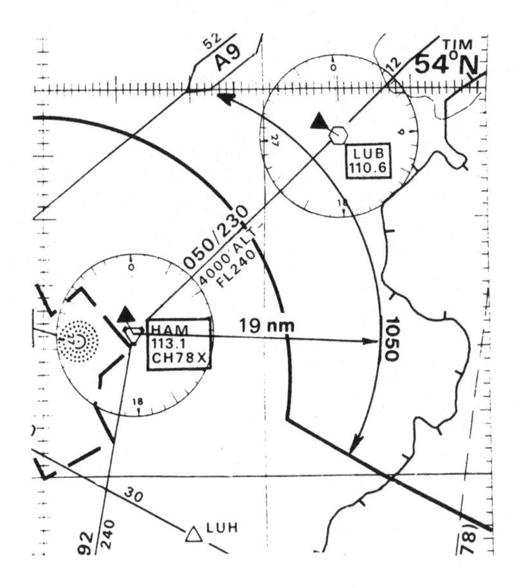

Fig. 2.12

magnetic variation is applied and then the magnetic bearing (QDR or radial) or its reciprocal (QDM) is transmitted to the aircraft either by electronic means so as to give a display on the cockpit indicator, as in the case of VOR, or by RTF to give the bearing to the pilot orally. In each case the procedure for reducing the bearing to obtain the direction to plot on the chart is the same. If it is not in the form of a magnetic bearing from the ground station (QDR or radial) but as its reciprocal (QDM), apply 180°. Then, having aligned the protractor with the magnetic meridian through the ground station, plot the QDR directly. In Fig. 2.13, a QDM of 135° obtained from VOR WSR has been converted, by adding 180°, to the equivalent QDR of 315°. The protractor is aligned with the magnetic meridian at WSR and the bearing of 315°(M) plotted. The compass rose at the VOR station was not used for measuring the angle because of the difficulty of reading it to the same accuracy as can be achieved with the protractor.

Plotting ADF Bearings

These are the bearings of the NDBs measured by the ADF in the aircraft (see volume 1). It is important to realise that, unlike VOR and VDF, the bearing measurement is actually done in the aircraft, i.e. at the aircraft meridian, but the resulting position line will be drawn from the meridian at the NDB position. G/Cs do not cut all meridians at the same angle and this will have to be taken into account in our procedure for dealing with these bearings. ADF bearings may be presented in the aircraft either as relative bearings (i.e. measured from the aircraft's nose) or on an RMI (Radio Magnetic Indicator) as magnetic bearings. The reading against the head of the needle on these is often referred as a QDM but this is not strictly correct: QDMs are reciprocals of bearings measured at the ground station meridian whereas these are measured at the aircraft's meridian.

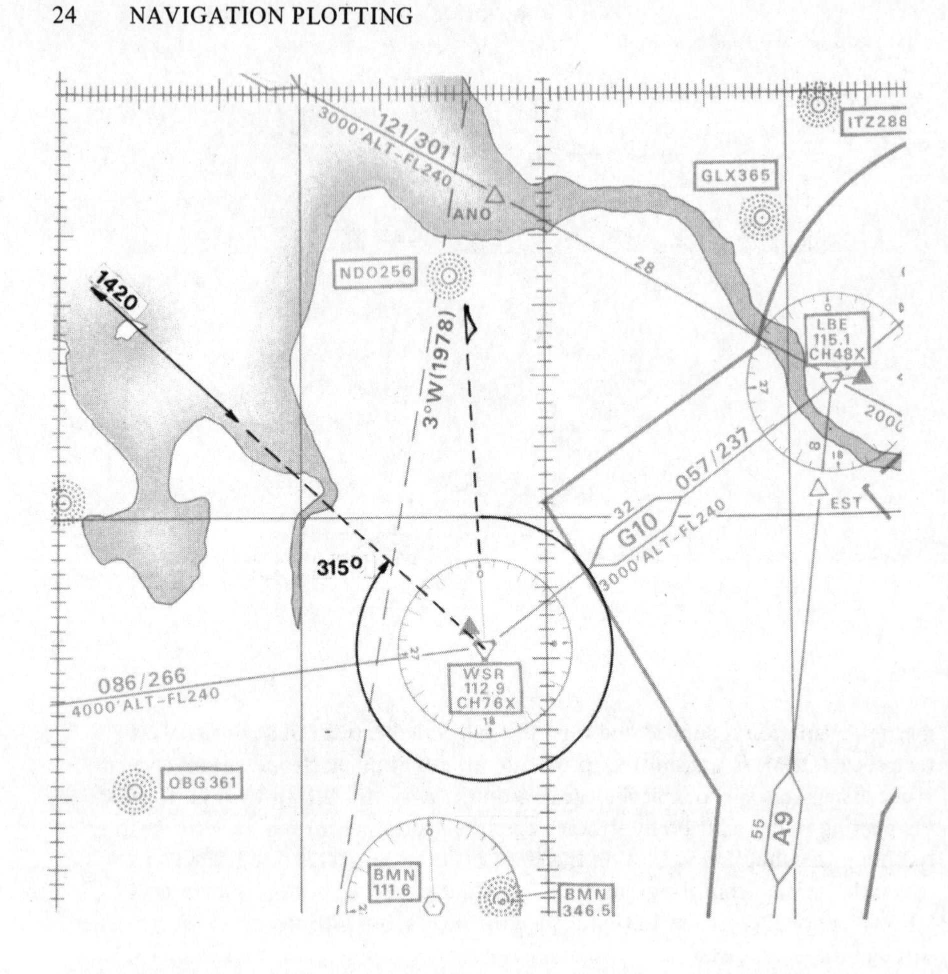

Fig. 2.13

In Fig. 2.14, an aircraft at a calculated (DR) position marked with a triangle, obtains an ADF bearing of 090°(M) of NDB LO on the RMI. As a double-check, the bearing is noted to be 080° on the Relative Bearing Indicator with the aircraft on a heading of 010°(M). Proceed as follows:

	RMI		Relative bearing indicator	
Bearing obtained	090(M)		080(rel)	
Aircraft varn.	−4	Heading	006(T)	010(M) −4
G/C bearing	086(T)		086(T)	
		086(T)		
Chart convergency (ch long $6\frac{1}{2}$E to $9\frac{1}{2}$E) x 0.8 = 3 x 0.8		$2\frac{1}{2}$	correction applied towards the Equator, i.e. towards 180°	
Reciprocal of bearing to plot		$088\frac{1}{2}$		
		±180°		
Plot from NDB		$268\frac{1}{2}$(T)		

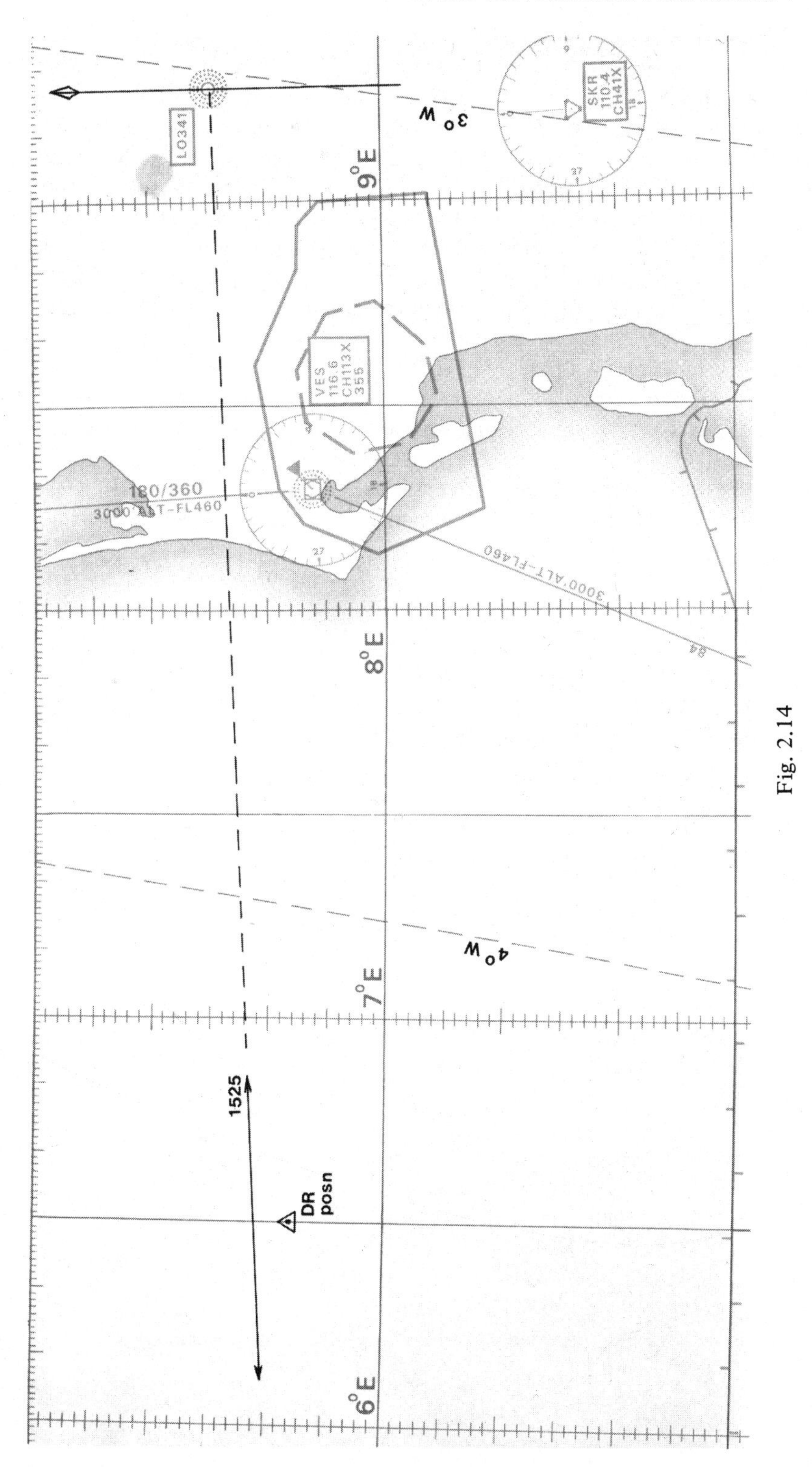
LO341
3°W
SKR
110.4
CH41X
9°E
VES
116.6
CH113X
355
180/360
3000'ALT-FL460
3000'ALT-FL460
8°E
4°W
7°E
1525
DR
posn
6°E

Fig. 2.14

For further details of convergency, see volume 3.

On this chart convergency is approximately 0.8° per degree of ch long. This indicates the change in direction of a G/C as it crosses successive meridians turning always towards the Equator. The convergency correction is always applied to bring the bearing measured towards the Equator (i.e. nearer to 180° in N latitudes) Having then formed the reciprocal, the bearing is plotted with the protractor aligned with the True meridian through the NDB – note that a small section of the True meridian has been drawn in to facilitate the alignment of the protractor. When the position line has been drawn in, it can be checked with the protractor that its direction in the vicinity of the DR position is indeed 086°.

Practical Exercises on the Chart

In the next chapter we deal with obtaining fixes using the position lines obtained at different times. Here, for practice purposes, you are asked to plot the bearings given and then read off the latitude and longitude of the intersection position. These would be referred to as simultaneous fixes. In each case the suggested correct answer is given in brackets. Your answer should normally be within 2′ of latitude and 4′ of longitude.

Question 1

1015 DR Position 54°N 03°E
SPY VOR (5232N 0451E) QDM 148
SPY DME Range 111 nm
(Bearing plotted 328(M) Fix 5401N 0301E)

Question 2

1120 DR Position 5430N 0600E
EEL VOR (5310N 0640E) QDM 166
DHE VOR (5411N 0755E) QDM 111
(Bearings plotted 346(M), 291(M) Fix 5430N 0555E)

Question 3

1245 DR position 54N 06E
WM NDB (5334N 0748E) bears 116 on RMI
LAK NDB (5231N 0534E) bears 194 on RMI
(Bearings plotted 293(T), 010(T) Fix 5401N 0602E)

Question 4

1415 DR Position 5430N 0230E Hdg 050(M)
WM NDB 064(rel)
LAK NDB 093(rel)
(Bearings plotted 292(T), 319(T) Fix 5448N 0209E)
The cut here is not very good (27°) and so even a slight inaccuracy in a bearing or in its plotting may produce several minutes difference in the answer.

Question 5

1520	DR Position 5400N 0230E	
	GV NDB (5206N 0415E) bears 157(M)	
	EEL VOR	QDM 112
	SPY DME	121 nm

(Bearings plotted 332(T), 292(M))

In this case, the three position lines do not meet at a point but form a triangle (cocked hat). In practice, it could well be decided that the VOR/DME information is more reliable than the NDB bearing and so the fix could be taken as the intersection of these two position lines (5355N 0226E). If all position lines are given equal weight, the centre of the cocked hat (strictly the intersection of the bisectors of the angles of the triangle) will be taken as the fix (5357N 0230E).

3: Practical Navigation Plotting

Equipment Required

All the examples in this chapter are based on the CAA Instructional Plotting Chart – Europe. It is recommended that each example is plotted as described. In addition to the chart, the following equipment will be needed:

navigation computer,
electronic calculator (not essential but useful),
protractor,
dividers,
1: 1 000 000 scale ruler,
pencil compasses,
HB pencil, sharpener and eraser.

Finding Initial Heading and ETA

1050GMT Overhead VES VOR (5536N 0818E) S/H (set heading)
EEL VOR (5310N 0640E) TAS 240 kt Forecast W/V 230/30 kt

What is the Hdg (M) required and the ETA?

Solution

From chart Tr(M) is 205° (printed along the route).
Applying 4°W variation the Magnetic W/V will be 234/30 kt and using the computer we can find: Hdg 208°(M), G/S 213 kt.

Distance to EEL again using charted data 84 + 73 = 157 nm

Using either the circular slide rule on the back of the computer (see computer handbook) or the electronic calculator:

157 ÷ 213 x 60 = 44 min

so ETA = 1050 + 44 = 1134 GMT.

Finding a Simple Simultaneous Fix

1114GMT VES VOR QDM 027
VES DME Range 83 nm

What is the aircraft's position at 1114?

Solution

VES QDM 027 + 180 = QDR 207

Plot 207 from VES with the protractor aligned with the magnetic meridian through the VOR beacon. To help in aligning the protractor it may be found helpful to extend the 000/180 points of the compass rose printed at VES.

Now draw in the range position circle of radius 83 nm and centre VES.

The intersection of the two P/L gives the fix position 5420N 0719E (answer should be within 2′ latitude and 3′ longitude).

Finding Tr and G/S W/V

Using the information in the preceding paragraph, deduce the average (mean) W/V that affected the aircraft since leaving VES, if the pilot steered 205(M).

Solution

Hdg(M) 205 TAS 240 kt TMG 207 (= QDR) and so the drift is 2° stbd.
The G/S is 83 nm in 24 min and so by slide rule or calculator:

$83 \div 24 \times 60 = 207$ kt

Setting heading 205 and TAS 240 kt on the computer, mark in the point where the drift line for 2° stbd cuts the speed arc for 207 kt (see computer handbook). The W/V can then be read off: 193M/34 kt (answer should be within 5° and 2 kt).

It will be noted that the magnetic heading was set against the True Heading index on the computer – as a result the W/V found is also Magnetic. The equivalent True W/V would be 189/34 kt.

This could be obtained either by using the True Heading of 201° on the computer or by applying the variation to 193°M wind direction already obtained.

Finding a Local W/V and Comparing it with the Mean W/V

In aircraft fitted with Doppler (see volume 1) or with on-board automatic navigational computers, it is likely that instantaneous values of Drift and G/S will be available. From these, the value of the local or spot W/V can be found using the same procedure as just described. The only difference will be the source of the information.

The question of which is the more accurate and which is the more useful W/V cannot be answered without a careful analysis of the circumstances. In its nature the Doppler-derived W/V will be a very accurate 'spot' W/V, but whether this is a good average W/V for the area would have to be a question of subjective judgement. The most useful W/V is that which is going to affect the aircraft in the future and who can say what this might be? The W/V to use is also, therefore, a question of personal judgement with due regard to the prevailing meteorological situation. It is worth remembering that even the professional met. forecasters can get it wrong so we must not be too disappointed if our estimates do not always work out.

Finding a DR Position and Revising Hdg and ETA

Using modern radio aids, an aircraft will usually be kept on the required track by a process of continual adjustment of the heading so as to maintain the correct TMG and Tr required readings on the indicators in front of him. The following procedure is not, therefore, likely to be encountered often. It could well be tested in a CAA examination question designed to ensure a candidate could handle the situation in the event of a systems failure.

Carrying on from the previous example – when the position at 1114 was established, you were required to calculate a new heading and ETA from 1117. It is quite common to 'DR ahead' for 3 or 6 minutes to give time to complete all the necessary calculations. Choosing 3 or 6 minutes (1/20 or 1/10 of an hour) simplifies the calculations.

Solution

Extend the TMG line between VES and the 1114 fix by a distance of 10 nm (3 min at G/S 207 kt). This gives the DR position (marked with a triangle) at 1117 as: 5410N 0712E.

From here the new required Tr to EEL of 198°(T) with a distance to go of 64 nm can be measured. Using the W/V found of 189°T/34 kt, the new Hdg required and corresponding G/S are: 197(T), 206 kt.

The revised flight time for 64 nm will be 19 min and so the revised ETA will be 1117 + 19 = 1136GMT.

The answers required are: Hdg 201(M), ETA 1136.

It must be emphasised that the solution will only be valid if the calculations are completed within the 3 minutes after the fix and if the W/V used is a fair estimate of what will be experienced until arrival at EEL VOR.

Finding DR Position at the Top of Climb

During climb, TAS and W/V will be continually changing and so it is not usually possible to establish an accurate top of climb position by calculation. In practice, if climbing away from a VOR/DME, there would be no need for the calculation as the actual position could be established so easily by a simultaneous fix. Similarly, if using an automatic on-board navigation facility, the aircraft position will be continuously available. In the absence of these facilities, calculating the DR position at the top of climb will have to rely on an estimated mean W/V – usually the forecast W/V for the mean altitude – and on an estimated TAS based on the planned mean RAS and the forecast temperature at the mean altitude. The procedure is best illustrated by an example:

1215GMT Position 5330N 0500E FL 100 COAT +4°C Climbing on Hdg 315(T). constant RAS 185 kt, mean forecast W/V 220/45 kt.
1245 Top of climb FL 240 COAT –24°C

What is the DR position of the top of climb?

Solution

Mean FL = $\frac{1}{2}(100 + 240) = \frac{1}{2} \times 340 = 170$

Mean temp = $\frac{1}{2}(+4 - 24) = \frac{1}{2} \times -20 = -10$

From circular slide rule (see handbook) Mean TAS = 244 kt.

From computer using Hdg, W/V and TAS, DR Tr and G/S: 325(T), 252 kt.

On chart draw Tr 325(T) from 1215 position for a distance of 126 nm (30 min at 252 kt) to establish the 1245 DR position: 5513N 0256E.

If your answer does not agree exactly, check that your mean Tr of 325(T) is correctly measured relative to the mean meridian of 4°E.

Finding the DR Position at the Top of Descent (TOD)

Modern aircraft perform better at higher levels and so it is advantageous to delay descending as long as possible – ideally the descent should be timed so that the aircraft arrives at the Terminal Radio Facility at just the required FL for the commencement of the approach procedures. In practice, such perfection is rarely achieved but calculations will often be done in the hope of achieving it. The procedure is very similar to that for the climb and will be illustrated by an example:

1415GMT Fix at SFR (5522N 0500E) on Tr to DHE (5411N 0755E) TAS 200 kt W/V 270/50 kt FL 170. Clearance received to descend so as to arrive at DHE at FL 50. Mean RAS for descent 150 kt, mean temp −15°C mean W/V 240/35 kt, rate of descent 800 ft per min.

Give the DR position and the latest time to commence descent, ETA at DHE and the Hdg(M) to steer on the descent.

Solution

Tr to DHE = 125°(T)
From computer Hdg(T) and G/S for level flight 133°, 239 kt
Descent mean FL $\frac{1}{2}$(170 + 50) = 110 mean TAS 175 kt
From computer Hdg(T) and G/S 135° 187kt
Descent time 12 000 feet at 800 ft per min 15 min
Descent distance 15 min at 187 kt 47 nm
On chart measure 47 nm back along the Tr from DHE to give the DR position for TOD 5438N 0648E
Distance from SFR to TOD 75 nm
At cruising speed 239 kt, time 19 min
ETA at TOD 1415 + 19 1434
ETA at DHE 1434 + 15 1449
Hdg(M) on descent 135 + 4 139°

Transferring P/Ls to Obtain a Fix

Situations may arise when, because of the shortage of aircraft equipment or ground facilities, it is not possible to obtain simultaneous fixes as we did in the last chapter. In these cases, it may be necessary to transfer position lines taken at different times so that they can be used together at a common time to obtain a 'running' fix. The basic technique requires a sensible assessment to be made of the aircraft's ground movement during the period of transfer and then to ensure that all points on the original position line are transferred accordingly. Consider the situation in Fig. 3.1, where three successive position lines have been obtained from one beacon using ADF and a relative bearing indicator (RBI). The aircraft is on a heading of 270°(T) and the best estimate of the TMG is 259°(T) with a G/S of 200 kt.

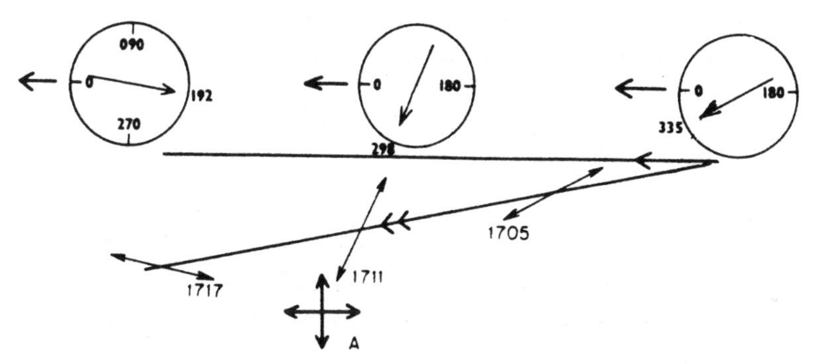

Fig. 3.1

Figure 3.2 illustrates the procedure for obtaining a 'running' fix. Assume the aircraft is actually at position A on the 1705 P/L – this is the point where it cuts the DR Tr line. Twelve minutes later at 1717, the aircraft would be at B. Distance AB = 12 min at 200 kt = 40 nm. Through B the transferred P/L is drawn parallel to the original P/L. The square navigational protractor with its rectangular grid will be found particularly useful for drawing parallel lines. It will be realised that every single point on the original P/L will have been moved to a corresponding point on the transferred P/L assuming that the aircraft movement is correctly represented in direction and distance by the vector AB.

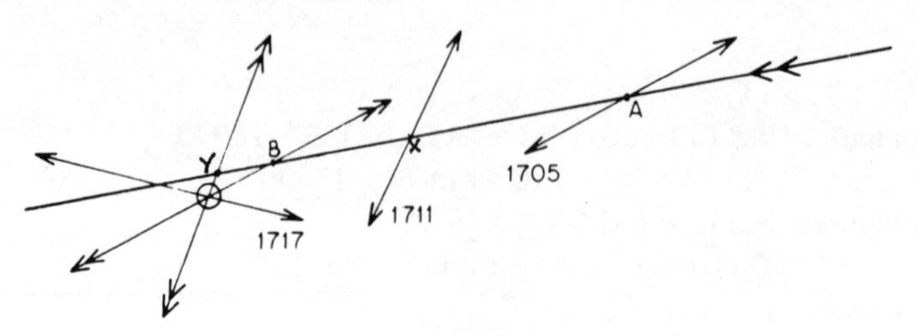

Fig. 3.2

Transferred P/Ls will suffer not only from any original inaccuracy but also additional inaccuracy due to the incorrect assessment of the aircraft's movement during the period of transfer. Of course, these errors could tend to cancel out but they could just as easily add together. In general, the longer the period of transfer, the more suspect the transferred P/L must be. It follows that periods of transfer should be kept to a minimum and the best possible assessment of the Tr and G/S prevailing during the transfer should be made. A Doppler Tr and G/S, if available, will be particularly useful.

Referring again to Fig. 3.2, which is not drawn to scale, the 1711 P/L is transferred in a similar manner. The point X being moved for 6 min at 200 kt = 20 nm to Y and the transferred P/L drawn through Y parallel to the original 1711 P/L. Note that original P/Ls are marked by single arrows at each end and the time they were obtained is written against them. Transferred P/Ls are marked with double arrows and no times.

The Cocked Hat

Figure 3.3 shows a situation which often arises in practice – 'a cocked hat'. This has already been referred to in the previous chapter. While cocked hats indicate that some error is present, it is not necessarily true that the size of the cocked hat indicates the magnitude of the error of the fix. With a consistent error present (e.g. compass deviation error in the case of ADF bearings), a fix obtained from bearings all on the same side of the aircraft as in Fig. 3.1 could give practically no cocked hat but the error would have moved all the P/Ls by approximately the same vector and so the apparently good fix will be in error by this amount.

If the bearings had been taken from beacons spread around the aircraft, ideally at 120° intervals, the constant error would have produced a larger cocked hat but its centre would be quite an accurate fix. The errors would tend to cancel each other out.

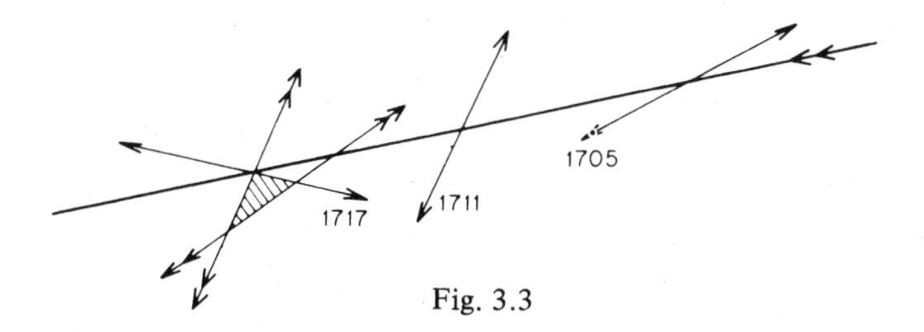

Fig. 3.3

Transferring Circular P/Ls

The principle here is exactly the same as for linear P/Ls. To ensure all points are transferred for the actual ground movement of the aircraft during the transfer period, we move the centre of the circular P/L (i.e. the DME beacon) as shown in Fig. 3.4 and then draw in the transferred P/L from this new centre. In practice, it would only be necessary to draw in the transferred P/L.

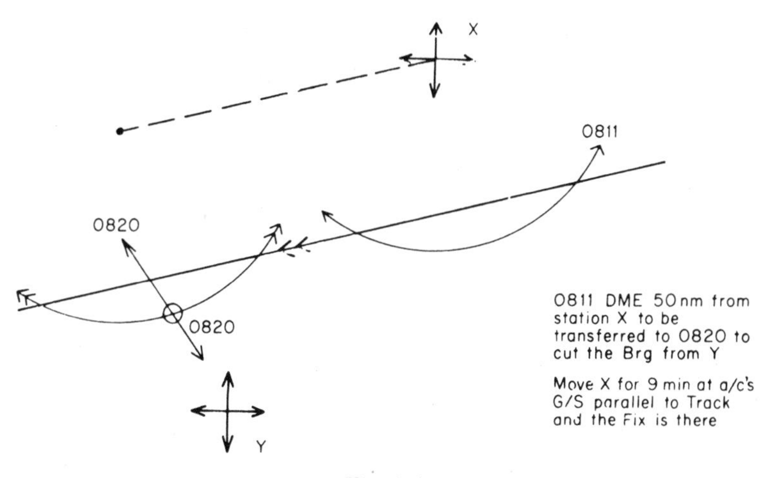

Fig. 3.4

Practical Example of a three P/L fix

Referring to Fig. 3.5, suppose the aircraft is in the vicinity of the Dutch coast around latitude 53°N and has been flying various headings and has now settled down on a Hdg of 085°(T) at a TAS of 180 kt with a forecast W/V of 000/30 kt. The following readings are then taken:

1215GMT	SPY DME range 25 nm
1218	ENK NDB bears 191(M)
1221	ENK NDB bears 219(M)

Solution

First use the computer to calculate the DR Tr and G/S 095°(T) 180 kt.

In the absence of any other information as to the aircraft's position, just draw in a track line of 095° anywhere in the general area. For transfer purposes we are only interested in the direction and distance of movement – the actual track location is irrelevant unless there was a question of deciding aircraft magnetic variation or the convergency values required with the NDB bearings. Through the SPY position draw in a transfer track of 095°(T) for a length of 6 min at 180 kt = 18 nm and then from this new centre describe the arc of the transferred P/L using a radius of 25 nm. Now calculate the bearings to plot:

Time	1218	1221	
Brg(M)	191	219	
a/c varn	−5	−5	May have to be revised if area assumed for the
Brg(T)	186	214	aircraft proves to be wrong
convergency	0	0	
	186	214	
take recip.	180	180	
plot (T)	006	034	

The 1218 P/L is now drawn in a direction of 006°(T) from ENK and the point where it cuts the assumed track (drawn in anywhere) is then moved along it for 3 min at 180 kt = 9 nm. The transferred P/L is then drawn through this point parallel to the original P/L.
Finally the 1221 P/L of 034°(T) is drawn in from ENK.
Position of the fix at 1221: 5256N 0532E

The Airplot
The procedures described up to now should be quite satisfactory for all normal navigation purposes but unusual situations can be visualised which would create problems for the methods just described. The airplot provides a very powerful solution which has the advantage of being comparatively simple to understand and use. An airplot, which is always started from a reliable fix is a graphical method of recording the true headings and air distances flown. In a still air situation, it would be a track plot. In the usual situation, the discrepancy, at any instant, between the airplot position and the actual ground position indicates the effect of the wind since the last fix, presuming all the information used is absolutely accurate.

In Fig. 3.6, if there had been no wind since the 1000 fix, the 1020 air position would also indicate the ground position at 1020. In this case, the aircraft is fixed at a position 244°/12 nm away from the air position. This indicates the effect of a W/V blowing away from a direction of 064°(T) at a speed of 12 nm in 20 min = 36 kt. Alternatively, if it was known that the W/V for the area was 064/36 kt, drawing in a wind effect (W/E) vector of 12 nm in a direction of 244° (i.e. away from 064°) would give a DR position at 1020. Of course, this could also be done by calculating the Tr and G/S for both the headings flown and then carrying out a track plot. The final result, if everything is perfect, should be exactly the same. In simple cases, involving only one heading, TAS and W/V, the Tr and G/S method is probably quicker and simpler but in multi-track cases the airplot will prove much simpler, particularly when a W/V has to be found.

Finding an Airplot W/V
The following example is illustrated in Fig. 3.7.

1200GMT	Fix 55°N 02°E Hdg 045(T) TAS 240 kt
1210	A/H (alter heading) 090(T)

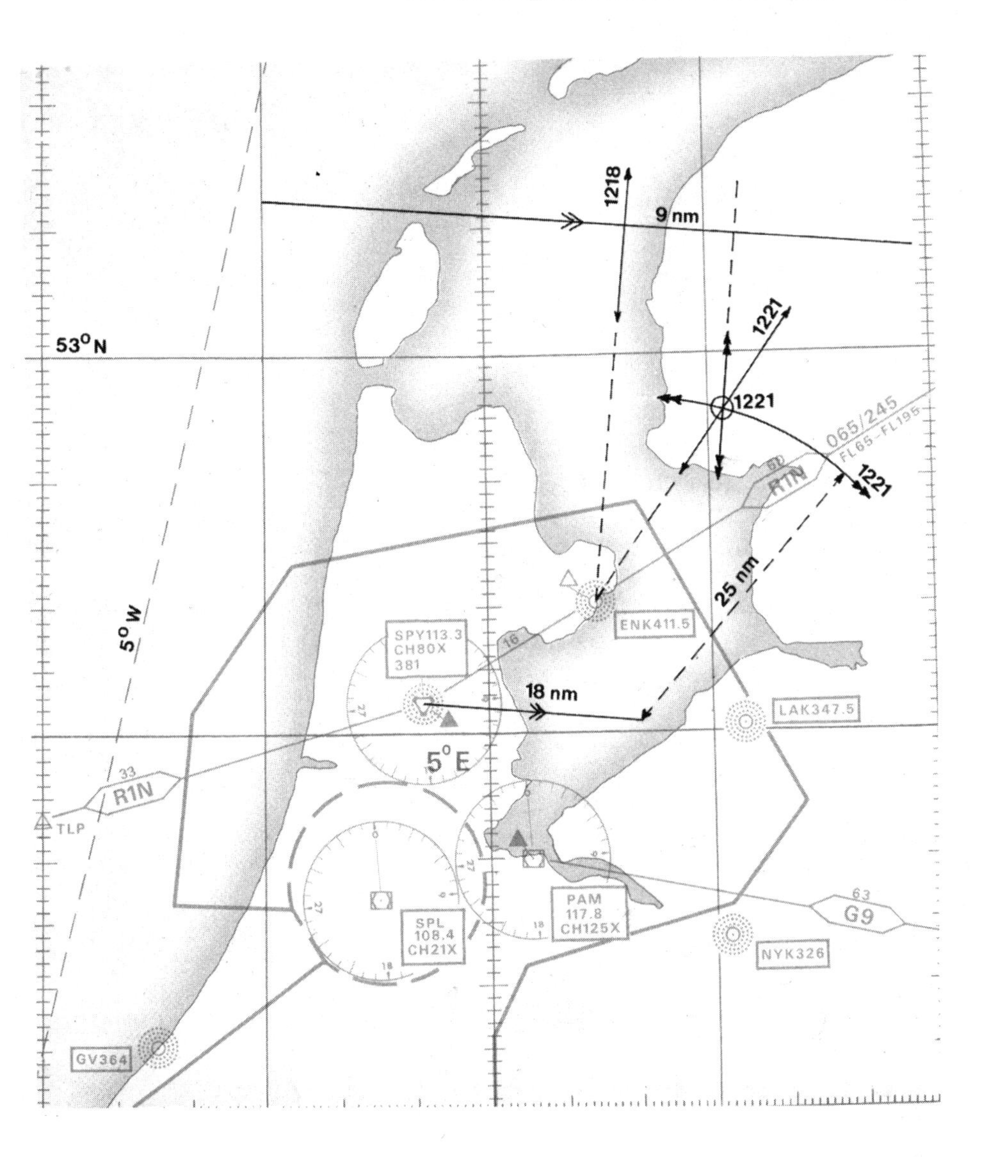

Fig. 3.5

1215	Reduce TAS to 210 kt
25	A/H 120(T)
33	Fix 5520N 0430E

Find the mean W/V from 1200 to 1233.

Figure 3.7 should be self-explanatory. It will be seen that at 1233 the aircraft's position indicates that the average effect of the W/V since the airplot started at 1200 has been to blow the aircraft 23 nm in 33 min in a direction of 283°(T), i.e.

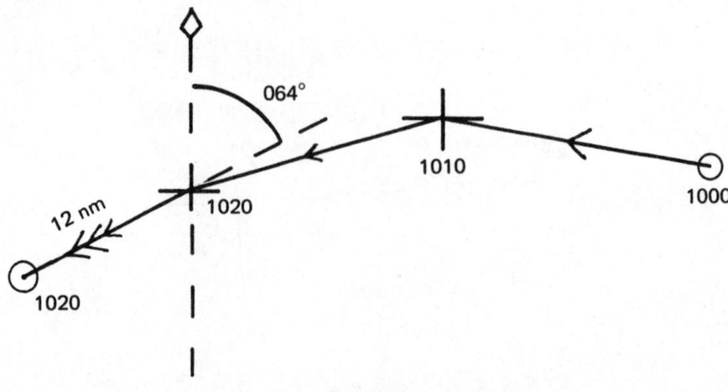

Fig. 3.6

away from a direction of 103°. The wind effect is therefore 103°/23 nm in 33 min and from this the W/V is calculated to be 103°/42 kt (23 ÷ 33 x 60 = 42). Note the single arrow vectors for the airplot and the three arrows for the W/E (wind effect) blowing from the 1233 air position to the 1233 fix position. Note also that the airplot involves a change of TAS and this is dealt with quite simply. Plotting the air position for 1215 when the TAS changed was not essential. The total air distance from 1210 to 1225 could be calculated:

(5 min at 240 kt) + (10 min at 210 kt) = 55 nm

and used to find the 1225 air position from the 1210 position.

Finding a DR Position by Airplot

Having found an average (mean) W/V by airplot, it may be required to 'DR ahead' i.e. to find a DR position for (say) 6 minutes after the latest fix. From this, having measured the new required track and distance to reach the destination, a new heading and ETA can be obtained. Referring to Fig. 3.7, the procedure to DR ahead for 6 minutes would be to start a new airplot from the 1233 fix by drawing in an airplot vector 120°(T) for 21 nm (6 min of TAS). From the resulting 1239 air position, a W/E vector is drawn parallel to the 1233 W/E vector and of length 4.2 nm (6 min at 42 kt). From the resulting DR position (5511N 0453E), the normal procedure of drawing in the Tr and distance required to reach a destination can be carried out. It is worth noting that, if this is done, and then further fixing and wind finding is required before reaching the destination, the airplot should be continued from the 1239 air position and NOT restarted from the 1239 DR position. If subsequent wind vectors are drawn in they will represent W/Es since the fix at 1233 NOT from 1239.

Practical Exercises

The following exercises are set in the form favoured by the CAA in their navigation papers. Having calculated the required answer and selected the nearest of the four choices offered, it is best to use the CAA answer for further calculations. For example, if you calculated that the TAS should be 182 kt and the nearest answer offered was 180 kt, it is recommended that you then use 180 kt for any subsequent calculations.

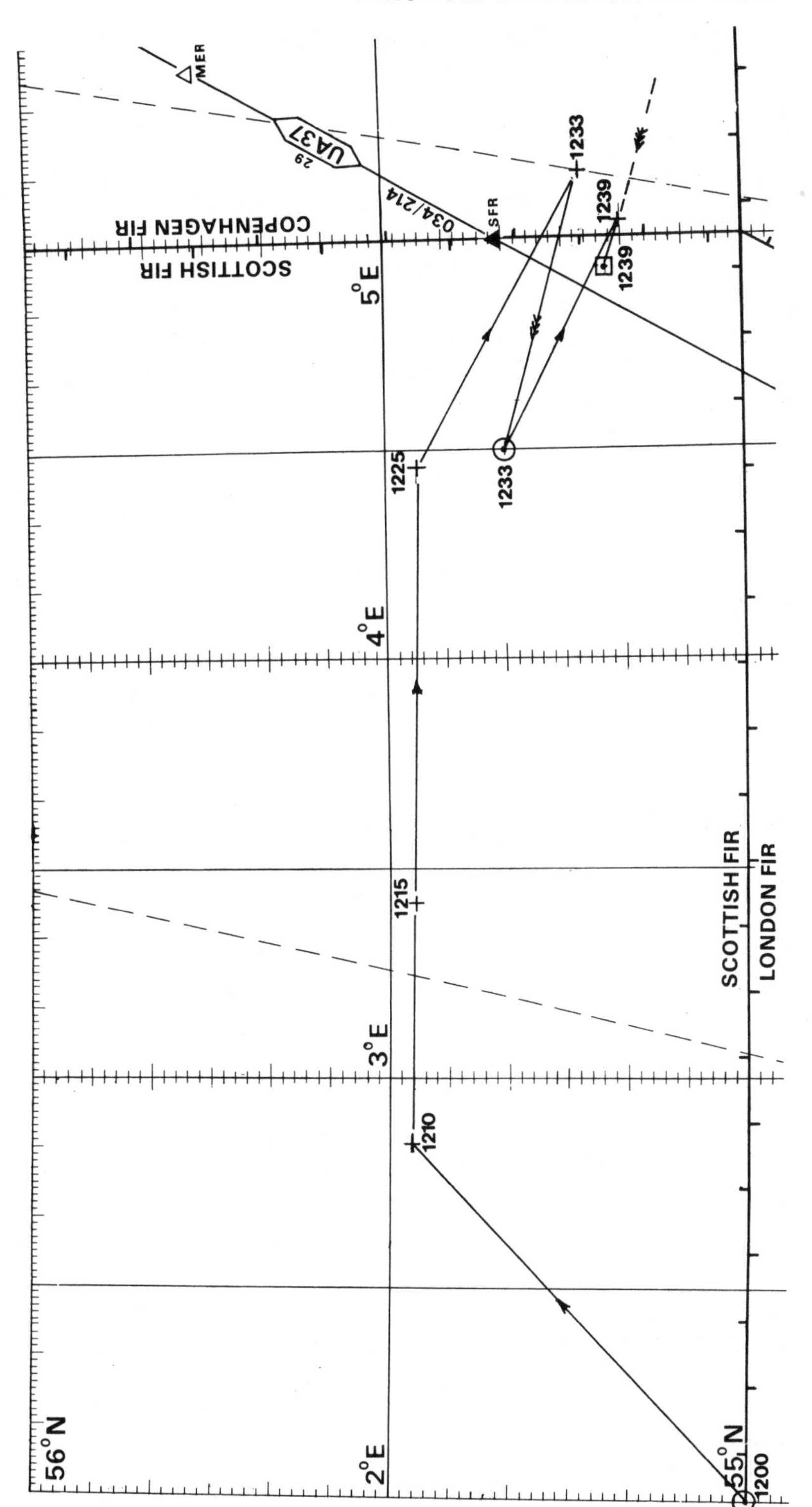
56°N
2°E
3°E
4°E
5°E
55°N
1200
1210
1215
1225
1233
1233
1239
1239
SCOTTISH FIR
LONDON FIR
SCOTTISH FIR
COPENHAGEN FIR
UA37
29
034/214
SFR
MER

Fig. 3.7

The following data relating to a flight should be used together with Instructional Plotting Chart–Europe to answer questions 1 to 9 inclusive.

1027 DR position (5430N 0330E), set heading for SPY VORTAC (5233N 0451E) FL 50, TAS 220 kt, forecast W/V 050°/30 kt.
You are cleared to join airway G9 at PAM VOR (5220N 0506E) at FL 90 and to commence the climb when overhead SPY VORTAC.
Assume for the climb: Mean TAS 150 kt.
Mean W/V 070°/55 kt.
1110 Overhead PAM VOR, FL 90, set heading for RKN VOR (5208N 0645E), W/V 130°/60 kt, RAS 200 kt, temperature −10°C.
1133 Overhead RKN VOR, alter heading for OSN VOR (5212N 0817E).
1139 Alter heading 10° to port.
1145 Heading 090°(M) Doppler drift 1°S G/S 188 kt.
1152 Overhead OSN VOR

Q1 The mean heading °(M) required at 1027 is:
(a) 170 (b) 146 (c) 156 (d) 160
Q2 The initial ETA for SPY VORTAC is:
(a) 1044 (b) 1100½ (c) 1059 (d) 1043
Q3 At SPY VOR the heading °(M) for PAM VOR is:
(a) 169 (b) 128 (c) 118 (d) 159
Q4 The rate of climb in feet/min required from SPY VOR is:
(a) 530 (b) 1 060 (c) 400 (d) 800
Q5 The cruising TAS from PAM VOR is:
(a) 235 (b) 232 (c) 223 (d) 226
Q6 The mean heading °(M) to steer at 1110 is:
(a) 116 (b) 098 (c) 112 (d) 104
Q7 At 1110 the ETA for RKN VOR is:
(a) 1130 (b) 1134 (c) 1132 (d) 1147
Q8 The mean W/V between 1133 and 1152 is:
(a) 098/48 (b) 088/37 (c) 120/74 (d) 098/70
Q9 The local W/V at 1145 is:
(a) 098/48 (b) 088/37 (c) 130/60 (d) 082/37

The following data should be used when answering questions 10 to 16 inclusive:

1707 SPL VOR/DME (5217N 0445E) RMI reads 098°, range 90 nm, FL 110, TAS 258 kt, W/V 250°/50 kt, heading 023°(T).
1728 GV NDB (5205N 0415E) RMI heads 178°
1732 NDO NDB (5347N 0849E) RMI reads 100°
1738 Alter heading SFR (5520N 0500E)

Q10 The position of the aircraft at 1707 is:
(a) 5251N 0217E (b) 5229N 0219E (c) 5221N 0343E
(d) 5221N 0217E
Q11 The theoretical range in nm of the SPL VOR/DME assuming its aerials are at sea level will be:
(a) 131 (b) 95 (c) 38 (d) 48

Q12 The DR Tr°(T) and G/S in kt at 1707 is:
(a) 015/288 (b) 023/293 (c) 030/297 (d) 030/293

Q13 Assuming a VHF range of 100 nm and a G/S of 300 kt, the aircraft will leave the coverage provided by the SPY VOR/DME at:
(a) 1724 (b) 1728 (c) 1726 (d) 1731

Q14 The position of the aircraft at 1732 is:
(a) 5404N 0408E (b) 5405N 0347E (c) 5423N 0341E
(d) 5355N 0410E

Q15 The mean W/V from 1707 to 1728 is:
(a) 250/50 (b) 146/56 (c) 326/56 (d) 259/56

Q16 The DR position at 1738 is:
(a) 5450N 0402E (b) 5418N 0440E (c) 5429N 0436E
(d) 5430N 0408E

The following data should be used when answering question 17:

1400 EEL VOR/DME (5310N 0640E) set heading 013°(T), TAS 190 kt on track to VES NDB/DME (5536N 0818E)
1435 Doppler drift 5°S G/S 210 kt
1440 HUU NDB (5428N 0905E) bears 121° relative
1440 VES DME (5536N 0818E) range 37 nm
1444 JEV NDB (5331N 0801E) bears 171° relative

Q17 The aircraft position is:
(a) 5458N 0813E (b) 5508N 0814E (c) 5511N 0822E
(d) 5513N 0815E

The following data should be used when answering questions 18 to 20 inclusive:

1705 Overhead EEL VOR/DME (5310N 0640E) set heading SPY VORTAC (5232N 0450E), TAS 160 kt, FL 80, W/V 220°/30 kt.
You are instructed to cross ENK NDB (5240N 0514E) at 2 000 feet.
Mean rate of descent 500 ft/min at constant RAS 120 kt, mean temperature + 10°C and mean W/V 250°/25 kt.

Q18 The G/S at 1705 is:
(a) 188 (b) 160 (c) 132 (d) 140

Q19 The DR position at the top of descent is:
(a) 5251N 0546E (b) 5251N 0516E (c) 5256N 0558E
(d) 5238N 0522E

Q20 The ETA at the top of descent is:
(a) 1717½ (b) 1722½ (c) 1719 (d) 1729½

4: Plotting on Other Charts

Up to now all plotting procedures have been done on the Lambert's Conformal Conic. This is quite logical as the chances are that any plotting required will be done on charts which are readily available on the flight deck. These are likely to be Radio Facility Charts and the majority of these are on the Lambert's Projection. It is just possible, however, that other charts may be encountered and this chapter will highlight any differences encountered in comparison with the Lambert's.

Chart Conformality (Orthomorphism)

Sensibly, any chart that is used for navigational plotting should be conformal. Volume 3 gives greater detail on this property, but here we should note that, unless a chart is conformal, measuring directions and distances on it will be extremely difficult if accurate results are required. Fortunately, the majority of charts encountered in civil aviation will be conformal (orthomorphic).

Practically every chart used will have its method of projection noted on it. Often the word conformal or orthomorphic will appear in the name. For your information the projections encountered are likely to be as follows:

Conformal	Not Conformal
Lambert's Conformal Conic	Gnomonic (Polar, Oblique, Equatorial)
(Standard) Mercator (cylindrical conformal)	
Transverse Mercator	Orthographic
Oblique Mercator	Equi-distant
Stereographic (Polar, Oblique, Equatorial)	

Classification of Charts for Plotting

For plotting purposes, all conformal charts can be placed into two categories:

Category M	(Standard) Mercator
Category L	All others – usually one of the following: Lambert's Conformal Oblique Mercator Transverse Mercator Polar Stereographic

Using Category L (for Lambert's)

The use of Lambert's has already been dealt with in Chapters 2 and 3. Using any of the other charts in this category will not cause any problems if similar procedures are adopted. It will be necessary to remember that the value of chart convergency varies from chart to chart as is explained in volume 3. In addition, certain problems will occur when flying in Polar regions but these can be overcome by using Grid Navigation techniques as described in the next chapter. Grid Navigation can be used outside Polar regions but inside these regions it is essential that it is used.

The Suitability of Category M Charts for Navigational Plotting

In the past, the standard Mercator chart has been used extensively for plotting. The RAF, for example, produced Mercator plotting charts covering the whole world apart from the Arctic and Antarctic. As is described in volume 3, the Mercator is an excellent chart for use in Equatorial regions (say 15°N to 15°S) but outside these areas, its use is rather impracticable and fraught with all sorts of problems. Measuring distances requires considerable care and straight lines on the chart represent R/Ls which are not the shortest routes for an aircraft to follow and are not the paths taken by radio signals. It is interesting to note that automatic on-board navigation systems direct aircraft along G/C routes, i.e. along curved lines on a Mercator chart. This is why one international airline issued their captains with non-conformal Oblique Gnomonic charts to cover their main routes. Although non-conformal, they had the one great virtue of accurately showing a G/C as a straight line, and so a line drawn between two points would precisely represent the route followed by an automatically navigated aircraft.

Using the Mercator in Equatorial Regions

It is here that the Mercator comes into its own. Everything is right. Consider the properties:

Scale: This is practically constant. In general, the use of scale rulers is quite satisfactory.

Great Circles: In equatorial regions these are almost identical with R/Ls and these are straight lines on the chart.

Graticule: This is rectangular and practically square which makes for great ease when reading off or plotting positions.

Radio bearings: These are G/Cs but in these areas almost identical with R/Ls, i.e. cutting all meridians at the same angle, so no corrections are required when plotting these bearings to allow for convergency.

In short, nothing could be easier. Use it like a Lambert's but forget about convergency corrections for the ADF/NDB position lines.

Using the Mercator in Other Regions

The best advice here is 'Don't'. If nothing else is available, the following points will need to be remembered:

G/C routes: Straight lines represent R/Ls so long routes will need to be broken down into a series of R/Ls which, taken together, will approximate to the G/C required.

Measuring distances: The scale in middle latitudes is already varying quite rapidly, e.g. from 54° to 55° latitude, the scale increases by $2\frac{1}{2}\%$! It is very necessary to measure distances very carefully at the correct appropriate latitude. Scale rulers must not be used.

Plotting radio bearings: The best procedure is to convert the G/C bearing measured, whether at the station (VDF/VOR) or at the aircraft (ADF), into the equivalent R/L bearing and this, or its reciprocal, can then be plotted without further problems.

Plotting Radio Bearings on Mercator Charts in Middle Latitudes

Two procedures needed to be recognised according to whether the bearing is being basically measured at the aircraft (ADF) or at the ground station (VDF/VOR). The following tabulations show the working sequence:

Type of bearing	VOR/VDF		ADF	
	QDM		RMI	RBI
	± 180		Brg(M)	Brg(rel)
	QDR		a/c varn	Hdg(T)
	Stn varn		G/C (T)	
	QTE		CA	
	CA		R/L (T)	
To plot	R/L (T)		± 180	
		To plot	R/L (T)	

Notice the two important differences between the methods:

(i) Station variation is used for VOR/VDF but aircraft variation for ADF.
(ii) CA is applied to the bearing measured at the a/c for ADF and to bearing measured at the station (QTE) for VDF/VOR.

In both cases the CA is applied to bring the bearing nearer to the Equator, i.e. nearer to 180° in North latitudes and nearer to 000° in South latitudes.

Obtaining Conversion Angle (CA)

This is dealt with in more detail in volume 3. Here, we are concerned with the practical methods of obtaining and using it.

CA is the difference between corresponding R/L and G/C bearings and can be obtained by:

(i) Calculation: $CA = \frac{1}{2}$ ch long x sine mean lat
(ii) ABAC scale (chart margin), see Fig. 4.1
(iii) Approximate factor:

lat	0		6		18		30		45		64		90
factor		0		0.1		0.2		0.3		0.4		0.5	

Example

DR position 5715N 0010E, NDB 5522N 0300W

Approximate mean lat 56°N, approximate ch long 3°

On calculator: 0.5 x 3 x 56 sin = 1.2°

On ABAC (Fig. 4.1): 3° ch long off the scale so use 6° and divide answer by 2 = 1.2°

Using factors: from the table, factor for 55° lat is 0.4°, CA = 0.4 x ch long (3) = 1.2°

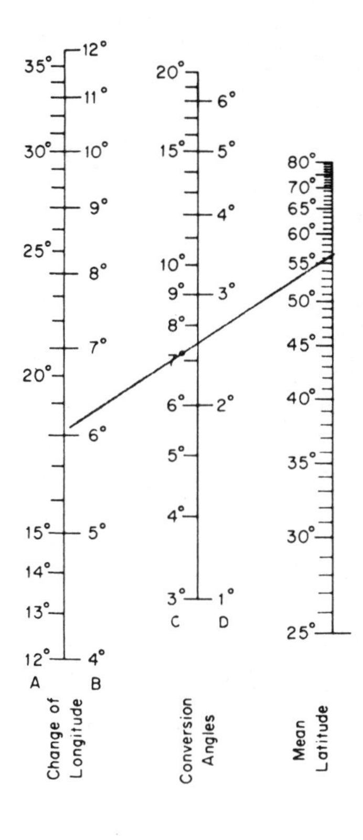

Fig. 4.1

Applying CA

CA should always be applied to the G/C bearing measured to bring it nearer to the Equator. For example, in the above example, if the aircraft had been on a heading of 330°(T) and a bearing of 267° had been obtained on the RBI of an NDB:

Brg (rel)	267
Hdg (T)	330
	597
	− 360
G/C Brg(T)	237
CA	− 1
R/L Brg(T)	236
	− 180
Plot	056 (T)

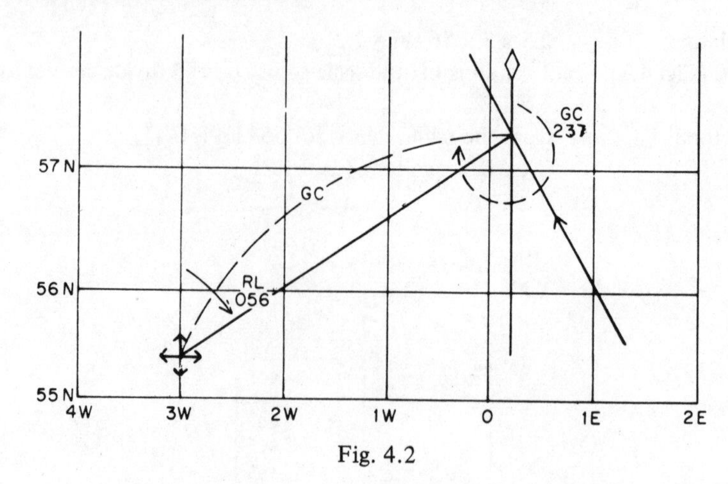

Fig. 4.2

Figure 4.2 demonstrates the principle – the dotted G/C shown is, of course, not a true representation. With a CA of only 1°, it would have been difficult to distinguish between the G/C and the R/L if they had been drawn in accurately. The diagram does demonstrate the correct sense of CA application, i.e. in this case it should be subtracted from 237° to bring the bearing nearer to 180°. If, in Fig. 4.2, the ground station was a VOR or VDF, the G/C measured would have been 055° and it would have been necessary to add the 1° CA to obtain the R/L of 056° to plot. Notice the application is still towards the Equator (180° in North latitudes).

Consider now a Southern hemisphere case as shown in Fig. 4.3. The G/C bearing measured at the aircraft position is 093° and the CA approximately 2° (4 x 0.4 = 1.6) and so applying the correction towards the Equator gives an R/L bearing of 091°. The reciprocal of 271° (T) is then plotted from the NDB.

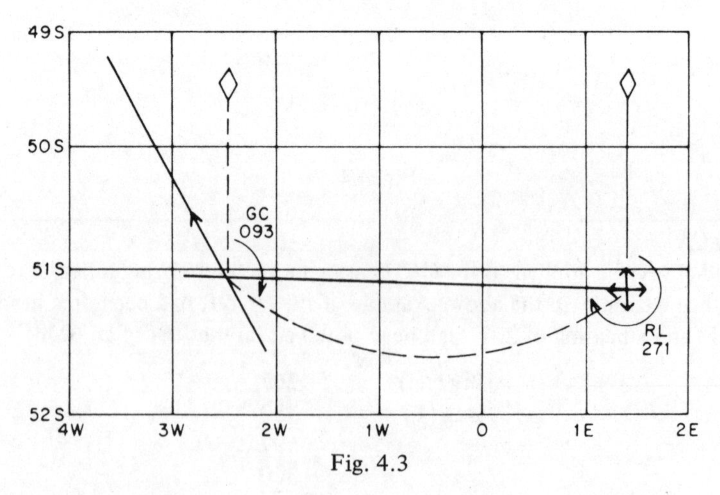

Fig. 4.3

Examination Questions

It is not likely that examination plotting on a Mercator will be required unless it happens to be in the Equatorial region. In this case, plotting is extremely simple as no angular corrections to bearings are required and a scale ruler can be used for distance measurement. The practice questions given here, therefore, will deal merely with the resolution of bearings to plot and not the actual chart work.

Questions 1-3

An aircraft is in DR position 60°N 05°E where the magnetic variation is 3°E. An RMI reading on an NDB at 60°N 20°E where variation is 5°E is obtained of 080°. Give the bearing to plot from the true meridian at the NDB on the following:

Q1 A Polar Stereographic where the chart convergency equals the ch long:
(a) 278 (b) 275 (c) 270 (d) 280

Q2 A Lambert's Conformal Conic on which the chart convergency for every degree of ch long is 0.8°:
(a) 278 (b) 275 (c) 270 (d) 277

Q3 A Mercator:
(a) 278 (b) 275 (c) 270 (d) 272

Question 4-6

The relative bearing of an NDB is measured as 235° by an aircraft on a heading of 065°(M) where the variation is 4°W. The ch long between the aircraft and the NDB is 11° and the mean latitude is 67°S. Give the true bearing to plot from the NDB on the following charts:

Q4 Mercator:
(a) 111 (b) 119 (c) 121 (d) 129

Q5 Polar Stereographic where chart convergency = ch long:
(a) 135 (b) 127 (c) 113 (d) 105

Q6 Lambert's Conformal where chart convergency equals 0.75° per degree of ch long:
(a) 108 (b) 116 (c) 124 (d) 132

Questions 7-9

An aircraft is in DR position 65°N 34°W where the variation is 30°W. RMI reading on a VOR in position 65°N 23°W where variation is 28°W, is 123°. Give the true bearing to plot from the VOR beacon on each of the following charts:

Q7 Mercator chart:
(a) 268 (b) 270 (c) 278 (d) 280

Q8 Polar Stereographic chart:
(a) 264 (b) 273 (c) 275 (d) 286

Q9 Lambert's chart on which chart convergency is 0.85° per degree of ch long:
(a) 266 (b) 273 (c) 275 (d) 284

5: Grid Navigation

Here we will touch only on the practical plotting on a Grid chart, as the theory of the system is discussed fully in the companion volume 3 'Navigation General'. The use of the Grid chart for air navigation is the easiest thing in the world, overprinted on any chart with converging meridians.

A reference meridian is paralleled across the chart in an outstanding colour, and this Grid Line is used to measure angles, ignoring the meridians, to obtain, for example, a Track (G). Also dotted across the chart are the Grid Variation isogonals, called lines of Grivation. Before you mutter any imprecations, the happy word is that:

Track (G) ± Grivation = Track (M) and the sign of Grivation is treated as for Variation. Similarly for a Heading, of course, so that you at once have the Heading (M) to steer, and all problems of angular measurement on a chart with converging meridians are avoided.

A Heading (G) will in fact differ from the Heading (T) at any meridian by the convergence between that meridian and the reference meridian. This convergence has been applied algebraically to the Variation to give Grivation so that:

Heading (G) ± Grivation = Heading (T) ± Variation.

When working on the Chart, then, all angles (including W/V) can be used quite satisfactorily in Grid.

Have a check on the following run down, on a bit of the gridded Lambert Chart, N. Atlantic (see Fig. 5.1): the reference meridian on this one is in fact the Greenwich meridian, and n for the sheet is given as 0·748819, another way of saying 0·75. Convergence thus becomes for the N. Atlantic Lambert:

ch long x 0·75

A 5000N 4000W to B 5500N 5000W

(i) Measured from the 45W meridian: Mean G/C Track 310
R/L Track 310

Convergence = ch long x n
= 10 x 0·75
= 8°

This gives the initial G/C Track of 314 (R/L 310 + $\frac{1}{2}$ Convergence)
and final G/C Track of 306 (R/L 310 − $\frac{1}{2}$ Convergence)

(ii) Now the mean G/C Track of 310, keeping to the 45W meridian, would be 340 (M).

The Grid Track there is 344 (G)

Grivation there is 4E

giving a Track of 340 (M), no different from the basic solution. The Grid can be used overall, holding the a/c on the mean G/C, avoiding the com-

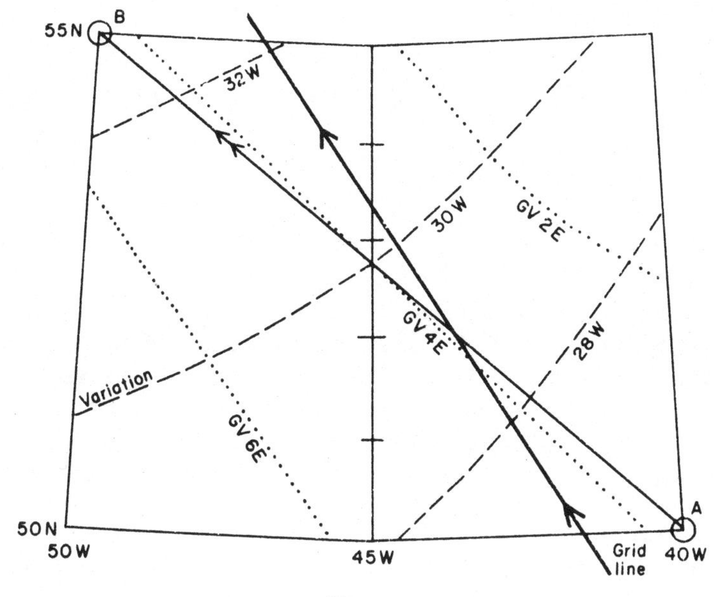

Fig. 5.1

plications of the converging meridians.

There must be no mixing, of course; all the information in Grid will evolve the correct navigational information. Take this example, on the computer:

Conventionally:	Mean G/C Track 310 (T), TAS 300 kt
	W/V 250/40
	∴ Hdg 303(T), Var 30W
	= Hdg 333(M) and G/S 280 kt
Grid:	Track 344(G), TAS 300 kt
	W/V 284/40(G)
	∴ Hdg 337(G), Griv 4E
	= Hdg 333(M) and G/S 280 kt

The relationship between Track(T) and Track(G) is the value of the convergence of the reference meridian or Grid line and the longitude in question; in this case, just a straight 45 x 0·75 = 34°, as the Grid line is the Greenwich meridian.

6: Radio Navigation Charts

The main producers of these are British Airways (AERAD) and Jeppesen. When one becomes familiar with one of these systems, it is not too difficult to transfer to using the other system. Every pilot has his own preference – usually for the first charts they learnt to use! Each has certain advantages and disadvantages but there is no clear winner. Both companies are constantly seeking to improve their products and ensure that the competition does not establish an unassailable advantage.

In this chapter, the AERAD charts will be used and it is quite important to obtain a recent edition of EUR 1/2 on which to follow out the examples quoted. Of course, it is possible that you will come across some discrepancies and, after careful checking to ensure that you are not reading the chart incorrectly, you should believe your more up-to-date edition. To maintain a high level of reliability, charts are re-published several times a year with often only very slight changes. It is, however, very dangerous to use out-of-date charts for operational purposes. Once a new edition of a chart is received, the previous one should be destroyed or clearly marked to show that it is no longer valid.

These charts are invaluable sources of information, not only on Radio Aids but also on all aspects of controlled airspace, restricted areas and radio communications. The charts are only part of the complete systems which incorporate manuals or supplements full of essential information regarding general aspects of aircraft operation plus local area, instrument approach charts, aerodrome plans and charts detailing SID (Standard Instrument Departures) and STAR (Standard Terminal Approach Routes). The complete system package will be accepted as an integral part of the aircraft Operations Manual which every public transport aircraft has to carry.

AERAD charts are mostly on Lambert's projection and are printed in grey, black and blue.

EUR 1/2

The following information is to be found in black:

Airway centrelines	Graticule figures
Radio facilities	Bearings and radials
Control zone boundaries	CTR, TMA, ATZ limits
FIR and ASR boundaries	Aerodromes listed in the Supplement

Under the blue colour look for variation (isogonal) lines at the top and bottom edges of the chart, danger, restricted and prohibited areas, aerodromes not listed in the Supplement, training and military areas, coastlines and lakes, and safe clearance altitudes within blocks of 1° latitude and 1° of longitude.

TMA and sub-TMA boundaries are in grey.

Now to learn to recognise things on the chart:

Controlled Airspace

Airspace left uncoloured on the chart (and this applies to other specification charts) – that is, all airspace shown in white – is controlled airspace. The rules regarding a flight in controlled airspace are fully dealt with in our companion volume *Aviation Law for Pilots.*

Airway

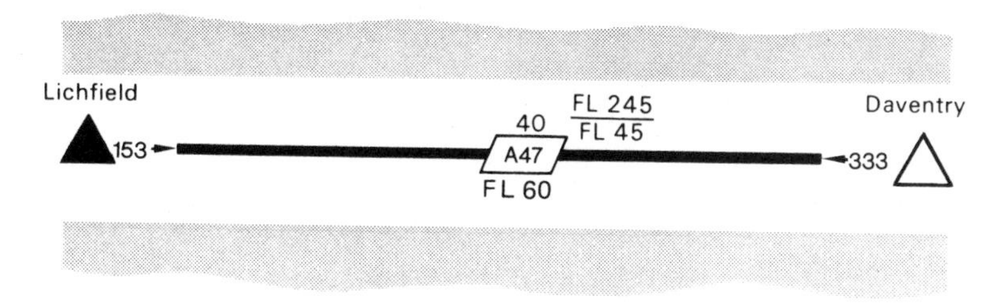

Fig. 6.1

1. Airway centre line is shown in thick black line, with the name of the airway in the centre. In Fig. 6.1 it is Amber 1E.
2. Immediately on top of the Airway name is the distance figure. This is the distance between two reporting points, compulsory or on request, the triangles at start and finish in Fig. 6.1. In this case the distance is 40 nm (all distances are in nautical miles) between Daventry and Lichfield. Daventry is shown as on request by an open triangle and Lichfield as a compulsory reporting point by a solid triangle. Facilities as shown on page 51 are often superimposed upon the triangles.

   Be careful when extracting distances. Distance breakdown occurs between reporting points and can happen between a reporting point and a sector point (X) on the route. If you have time, open up your dividers between two points and check the figure against the nautical mile scale (and not, please not, against kilometre) given on top of the chart. If after a rough check your distance agrees reasonably with the printed distance, take printed distance. If you have made a careful check and the printed distance agrees within 2–3 miles, take the printed distance.

   However, the printers do make mistakes and if your measured distance disagrees with the printed distance beyond the above limits, take the measured distance, but do point out in your answers in an examination why you are using measured distance.
3. Tracks

   Track angles are given at the beginning, immediately after the facility/reporting point. These tracks are Magnetic. In fact, no tracks, headings, bearings or radials on this (or any other Aerad) chart are True – they are all Magnetic. A one-way airway would have the track shown on one end only: the airway designation box indicates that the airway is one way, thus:

R3

4. Minimum Flight Level
The minimum flight level available on that particular sector of the airway is given immediately below the Airway designation: FL 60 in Fig. 6.1. This is the lowest level you can apply and get clearance for flight on this sector.
In mountainous territory it would be too risky to give a minimum flight level. Here, it will be given in terms of altitude, e.g. 14 000. Examples occur in the Zurich area, and the pilot merely flies odd or even thousands of feet on QNH, i.e. with the regional barometric setting given by Control; this will read altitude above mean sea level. (The full story is given in the altimeter chapter in volume 3.) Alternatively, the minimum flight limit may be defined in terms of both, FL and altitude, e.g.

FL 50
(Min. Alt. 4 500)

This simply means that FL 50 is available for flight provided it at least equates 4 500 feet on QNH.

5. Airway Vertical Limits
Airway vertical limits must not be confused with minimum flight levels. These vertical limits legally define the controlled airspace forming an airway, CTR, TMA, etc. In our illustration the airway out from Daventry stretches from FL 45 (the base of the Airway) to FL 245 (the ceiling of the Airway–Upper ATS routes start at FL 250). If these limits were to change, a pecked line at right angles to the airway centre line would indicate where the change takes place, and the ceiling and base limits would be printed on each side of this dividing line. This lowering of the base does not affect the minimum flight level available for flight on the sector.
The easiest way to distinguish between the two is: where FL or Altitude figure stands on its own on the chart it is the lowest limit at which the flight can be made; where FL or Alt appear in the form of a fraction, then those are the vertical limits imposed.
It may be mentioned that the base of an airway, or TMA need not be in terms of FL, it may be given as altitude as well, e.g. the vertical limits of Red 123 SW of Brookmans Park are

$$\frac{\text{FL } 245}{3\ 500}$$

Lastly, in the U.K. there is at least a 500 ft clearance between the base of the airway and the lowest aircraft on the airway.

6. Safe Clearance Altitude
The method of showing this along the airway in thousands of feet (3.3 = 3 300 ft) is being replaced by the area system. Every 'box' of 1° of latitude and 1° of longitude on the chart will contain figures in large blue type:

34

indicating a safe clearance altitude for the 'box' of 3400 ft. The figure incorporates the following clearances:

Terrain up to (1 000 ft)	5	10	15	20	over 20
Clearance (1 000 ft)	1	1.5	2	2.5	3.0

It should be emphasised that a Commercial Pilot is required to use the clearances laid down in his Company's Operations Manual which might exceed these.

That's just about all as far as airway symbols are concerned. We must warn you that the layout shown above (Fig. 6.1) is the ideal layout. In congested areas, information may be scattered all over the place. Now for the remaining symbols that you will see on the chart.

Reporting Points

Small triangles, as shown above at Daventry and Lichfield. A full black triangle (complete block) is a compulsory reporting point and you must report there unless the current official advice is to do otherwise. A hollow black triangle is an 'On Request' reporting point where ATC may request you to report.

Facilities

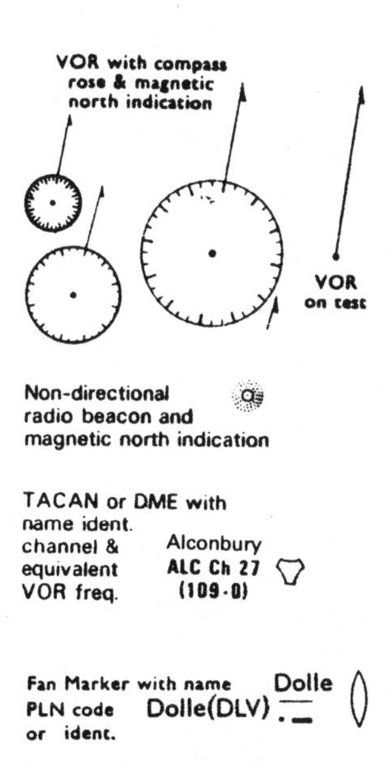

Fig. 6.2

NDBs. Those tiny black flags above NDB symbols (but some distance away) are not indicative of holes at the local golf course. The flag direction is the direction of Magnetic North at the NDB below it. The flag to the west of the vertical line indicates westerly variation. In Fig. 6.2, the variation is easterly.

VORs. The direction of Magnetic North is indicated by a line with a flag at its top, from the 000° radial. The convention regarding easterly/westerly variation applies.

Note that a VOR on test has a special symbol (circle without calibration points) whereas an NDB on test has no separate symbol. In practice, an NDB is not inserted on the chart until it becomes fully operational.

TACAN/DME. Although ICAO has separated the two, a single symbol is used on Aerads to indicate both. The separation is achieved in the information box, see below.

Fan Marker. Some of them are still left. You require a 75 MHz receiver to get the signals.

Facility Frequencies

Against each facility symbol, the call sign and frequencies of the radio facilities are given. 'SND' 362.5 for example is SOUTHEND NDB on frequency 362.5 kHz, call sign SND. In no time flat after some experience you will recognise VORs, NDBs, TACANs and so on, just from the frequencies given. On actual Airways, the information is enclosed in a box. SND has no official route going through it.

Another point to note is that the callsign 'SND' is in quotes, as is practically every other NDB in the UK. This indicates that the emission is NON A2A (or A2 as it used to be called). As explained in volume 1, the signals of this type can be heard without switching to BFO or CW on the ADF. Omission of the quotes indicates that the emission is NON A1A (or A1) for which it will be necessary to switch to CW/BFO to hear the callsign. NDBs of this latter type will still be found in many parts of the world.

As for DME and TACAN let us take a look at the following frequencies noted on the chart.

Pole Hill	Coningsby	Camelot
POL 112.1	CGY Ch 48	CMT 113.5
Ch 58(DME)	(111.1)	CMZ Ch 82 (DME)

You will find the first and second of these on Fig 6.6. Pole Hill is a typical airways VOR/DME using the standard frequency pairing. Tuning the VOR to 112.1 MHz will automatically set the DME to Channel 58 and the pilot will have a continuous read-out of bearing to and distance from Pole Hill. Coningsby is a military TACAN from which ranges can be obtained by tuning the VOR to 111.1 MHz and so automatically selecting Channel 48 on the DME. Bearings will not, however, be obtainable. Camelot is typical of a DME which is not co-located with the VOR. This is indicated by the DME having a different callsign ending in Z. In such cases they are fairly close together (say 1 nm apart) and frequency paired so that they can be used en route to obtain bearing and distance fixes.

Control Zones, Special Rules Zones/Areas, Military ATZs

CTR (Control Zone), CTA (Control Area), TMA (Terminal CTA), SRA (Special Rules Area), SRZ (Special Rules Zone) and (Mil)ATZ (Military Aerodrome Traffic

Zone) are all shown, apart from the (Mil) ATZ, as white areas outlined by black pecked lines. The vertical limits are shown:

Honington (Mil) CTR	Scampton (Mil) ATZ	Ramstein (see (Fig. 6.8) TMA A
4000	3460	FL 245
g	g	1000g

g = ground or above ground level; unqualified numbers are altitudes.

Altimeter Setting Regions (ASRs)

These are shown by longish pecked lines in black, the lines defining the boundaries. The names of the regions appear somewhere along the pecked lines, and when you cross this line, you should change your altimeter setting to the QNH value of the region you are entering if you wish to check for terrain clearance. For normal en-route flying, the standard altimeter setting of 1013 mb will be used.

FIR Boundaries

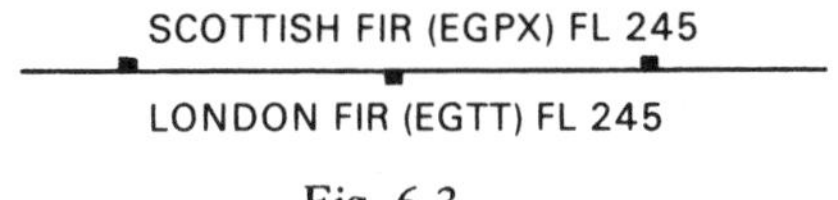

Fig. 6.3

The name of the FIR, its ICAO four-letter code, and the upper limit of the FIR which in each case is FL 245 is given. It is worth mentioning that the FIR does not have to be a straight line. FIR boundaries quite often coincide with national frontiers. On Fig. 6.8 find and note the shape of the boundary between the PARIS and FRANKFURT FIRs.

Bearings and Radials

These generally define reporting points, where they are not co-located with radio facilities; in the main they are on-request points. There are a few left, but now they are invariably radials, as illustrated in Fig. 6.4. If you do come across an odd NDB bearing, follow the line down to the NDB itself where the bearing figure is inserted.

In Fig. 6.4 reporting point LOGAN on R1 is defined by a VOR radial. The presentation is standard on all Aerad charts: 019°R simply means it is on 019° radial from VOR that is traced at the other end of the bearing line. In this particular case since Dover VOR is not on this side of the chart, its reference is added here in terms of its callsign and frequency.

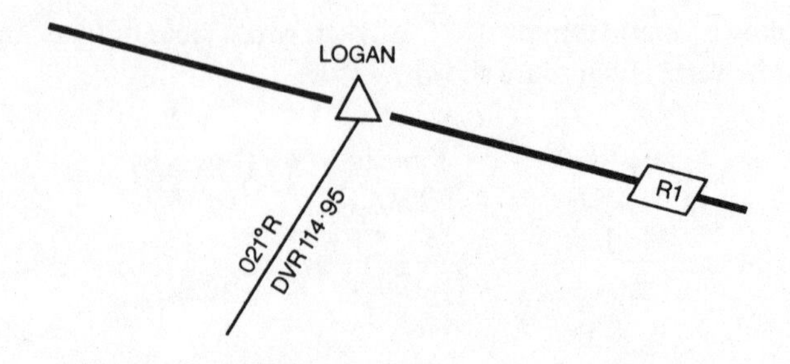

Fig. 6.4

Advisory Routes (ADRs)

Designation of all advisory routes up to FL 245 is by the letter D as first letter, e.g. DR1(Lower ADR Red One). These will be recognised on the chart by grey areas (grey because ADRs are uncontrolled) delimited by white outlines on either side of the centre line. The practical aspect of flight on ADRs is fully investigated in *Aviation Law for Pilots*.

You will find, out of Ottringham, for example, a series of routes over the North Sea simply designated by a rectangle containing the letters DT for direct track.

DT

These are commonly used routes over the uncongested areas. The distances have been calculated by Aeradio themselves, and they are therefore unofficial as it were, but they like to be helpful.

Practically all airways and all ADRs are 10 nm wide (5 nm either side of the centre line) in Great Britain.

In blue you will see the outlines of restricted airspace (Danger, Restricted and Prohibited areas), isogonals and graticule outlines. You will need to search the edges of the chart for the short sections of isogonal that are shown.

Danger Areas

Danger Areas should be checked before flight; the chart shows permanent Danger/ Restricted areas as a continuous blue line; temporary ones (that is, those activated by Notam) are shown by small pecked lines. Prohibited areas are filled in with tiny blue dots.

On the last fold on each side of the chart you will find the Airspace Restriction Panel which sorts out Danger and Restricted areas (which otherwise look alike on the chart: continuous line if permanent or during published hours, broken if temporary) and give pertinent details about those and prohibited areas. It is necessary that you are able to decode printers' shorthand. Here are a few examples, taken from 'France'.

R1 820–2 600 g HN M–F

R1 is the ident of the Area, and the area concerned is a restricted area (R). Limits 820–2 600 g imply the figures are on QFE (above the ground), and it is restricted to Night hours M–F only. Quite a biggie, this one, if you trace it round.

R77 FL60 Permanent
Here the limit is from ground or sea whichever happens to occupy the area up to FL 60 (on 1 013·2 setting).

D43 FL 30–55 HJ (VMC) HN and Notified
This is a Danger Area (D), limits are obvious but under the remarks column we have been given times of operation: daylight hours if VMC, at night or when notified in Notam.

If the limit was given, say, as 1 000–FL 55, it means both limits on 1 013·2. 1 000 above ground is given as 1 000 g. Other abbreviations used are:
Unltd – unlimited.
Wkd – weekdays, Monday to Saturday (inc.).
M is Monday, Tu is Tuesday, and so on.
HJ daylight hours (proof of the mapmakers' deep knowledge of the French tongue).
HN – night.
agl – above ground level.
(S) Summer (W) Winter.

TMA Boundaries
These are perhaps the most difficult ones to spot. Main boundary and all sub-divisions of the TMA within it are shown by white areas (controlled airspace) bounded by thickish grey lines. What makes it difficult to trace is the fact that these boundaries generally run along the outer boundaries of Airways – just where the white colour finishes and the grey starts – and the lines seem to merge with the background grey. However, with a little practice you will soon be able to keep it in sight. Try Scottish TMA for a starter on your own copy of EUR/1.

Isogonals
These are shown as blue lines. The data and annual change are noted just under the map reference EUR/1.

Miscellaneous
1. Authorised routes (which are neither airways nor ADRs) are shown by black thin lines, with track and arrow at both ends if the route is two-way; if it is one-way, the route line ends with an arrow.
2. Quite a few States publish Flight Plan codes for reporting points. Where these are available, the pertinent letters in the name of the reporting point are underlined, thus: Matching.
3. If a reporting point is abeam a facility, and this is a part of its definition, it would be shown thus:

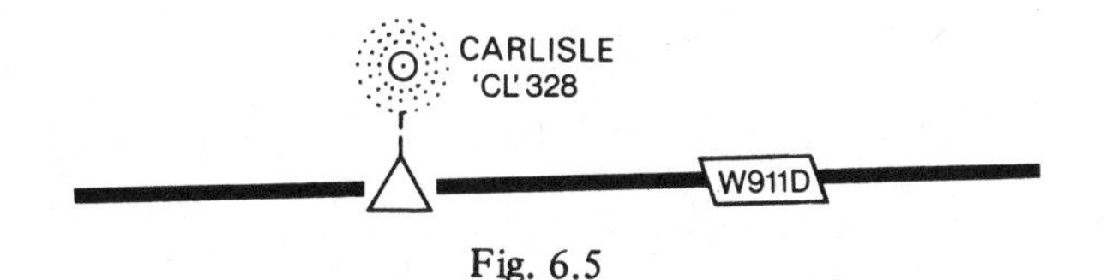

Fig. 6.5

Such a reporting point would be called 'Abeam' Carlisle and don't forget if you have got 8° starboard drift when you are nicely on Track lined up with departure or destination radials, then you're abeam CARLISLE when the relative bearing is 278 going Easterly and 098 going Westerly.

4. Airway Frequencies

For UK and W. Germany, frequencies for use on the airways are no longer shown on the chart. If required, they will be found in the appropriate AERAD Supplement.

On EUR/2 you will notice that countries on the continent sectorise the territory into convenient communication sectors, and the frequency for use in a particular sector is given in a prominent (so they say) place inside the sector (boxed in a parallelogram).

Paris Ctl	Brussels Ctl	
(NW of Paris)	(W of Brussels)	
128.2	128.8	128.45

Paris Control and Brussels Control are the callsigns.

More frequencies are listed on the front fold of the chart – CTR, FIS, and so forth. Have a look at them.

5. Flight Levels

Generally semi-circular rules apply on airways, and Aerad inserts 'odd' and 'even' when necessary on the airways (e.g. A25 at Dean Cross) although these are sometimes left out in cluttered areas. In the absence of these, fly odd thousands when Tr(M) lies between 000–179; fly evens when Tr(M) between 180–359. This is, however, only a general rule and many exceptions occur (perhaps you remember that rules of quadrantal and semi-circular flying do not apply in controlled airspace in IFR: you fly the level given you).

6. Coastal NDBs

Some of these NDBs do not provide continuous service.

Comer
'CM' 287·3
H + 04 & Ev. 6 mins

This NDB only transmits short bursts of CM in morse for 1 minute at 4 minutes past the hour and repeats every 6 minutes thereafter.

Information not on EUR1/2

Any information on routes at FL 250 and above is contained in High Level Chart H108/109.

Information on holding (holding point, pattern, time and min. alt.) is contained in a panel on 'area charts'.

The remaining symbols on EUR 1/2 are

Civil A/D   Civil & Mil. A/D   Mil. A/D   all with names

ADIZ (Air Defence Identification Zone) - in blue

Notes on Airway Flying

1. To stress the point again, distances are given between compulsory or on-request reporting points, the solid or open black triangles, most of the time, also between intersection or turning points marked with a cross, so watch it.
2. A report is not necessarily required on crossing a boundary from one Information Region to another, but you must be aware you have crossed it for your subsequent reports. The parallel of latitude 55N from 0500E to 0530W is marked with a pecked line indicating such a boundary, and the name of the appropriate FIR region set on either side of it, London & Scottish. Three boundaries meet on the 0530W meridian at 5355N, for example, easy to recognise here, but on B29 on 294M out of Nicky near Brussels the boundary is a winding river which is crossed 4 times in about 30 nm.
3. A report is made only to the Air Traffic Control Centre (ATCC) of the Flight Information Region (FIR) in which the aircraft is flying: thus, a report in the London FIR is to London Airways.
4. The message content is crisp and its form invariable, once communication has been established (Scottish Airways, this is Golf Alpha Sierra Tango on 124·9, do you read?). 'Scottish Airways, this is Golf Alpha Sierra Tango, Lichfield 27, Flight Level 80, estimate Oldham 39, over'. That's all. Who you are, where you are, your Flight Level, ETA next point, over, and memorise the form.

Let us consider a stage of a flight from HEATHROW to PRESTWICK via Blue 4 from the time of passing over BROOKMANS PARK (5145N 0006W) part of which is shown in Fig. 6.6 from EUR 1.

BROOKMANS PARK is a non-compulsory reporting point with a VOR/DME, callsign BPK and frequency 117.5 Channel 122. The route is clearly Westbound (334M) and so in the absence of any contra-indication, we will fly at an 'even' FL. The lowest acceptable is shown on the chart as 110 and so we will choose FL 120. We will be flying under the instructions of London Control and in accordance with the clearance they issued in response to the Flight Plan we would have submitted before take-off.

BEDFO is the next reporting point (on request only), it is 33 nm from BPK on a Tr 334(M). If arrival here is to be confirmed, the most accurate cross-check would be using DAVENTRY VOR/DME, QDM 270, range 20 nm (dividers or a suitable ruler will be needed to check this). These readings should be obtained simply by tuning 116.4 MHz on the VOR and checking the c/s DTY is being received before accepting the readings.

Another question that may be posed concerns the readings that would be expected on the BPK VOR. If the aircraft is fitted with an RMI, the answer is very simple. The head of the needle would indicate the QDM back to BPK, i.e. 154, or conversely the tail of the needle would indicate the QDR 334 – always presuming that you are accurately on track! If the aircraft is only fitted with a Relative Bearing Indicator (RBI) the tail of the needle should indicate the actual drift being experienced, i.e. 10°S would show as 010°(rel) or 10°P as 350°(rel).

The next compulsory reporting point is POLE HILL VOR/DME. Flying at FL 120 the reception range of VHF signals will be approximately $12.5\sqrt{(120)}$ =

137 nm approximately. Referring to the chart, the total distance from BPK to POL is 33 + 52 + 55 = 140 nm. It follows that POL could be used for most of the flight to give positive checks on the aircraft's position in terms of QDM and distance to go.

How clear are we of high ground? Flying on an airway in the UK always guarantees a clearance of at least 1500 ft above the highest obstruction within 15 nm of the centre line of the airway but apart from this, we can check the minimum clearance altitudes for the 1° lat/1° long 'boxes' along the route. These give altitudes of 2 200, 2 600 and 3 600 feet as far as POL.

Shortly before arriving at POL, the new Tr(M) required from POL of 345(M) would be selected for automatic flight or, if flying manually, it would be set on the Omni-bearing Selector (OBS). Arrival at POL would be indicated by the DME reading about 2 nm – at 12 000 ft, the aircraft would be about this distance above the ground beacon. Having passed POL, the VOR indications will reverse. On the RMI, the head of the needle will now show QDM 165° and on the L/R indicator, the word FROM will appear in the window of the indicator instead of TO.

A word about other controlled airspace on this route. The route starts in the London TMA, passes over the Bedford (Mil)ATZ, the Daventry CTA (not apparent on the chart) and the Nottingham (E Midlands) CTR (g to FL 75) and then goes through the Manchester TMA (up to FL 245). The actual positions of entry and exit are not easily identified on this involved area of the chart but this need not concern the pilot too much. Having been cleared to fly on Blue 4, he will be instructed when to change from one controller to another. Although it cannot be deduced from this chart, an aircraft flying below FL 155 will be controlled by Manchester from abeam Birmingham (about 5225N) until Shapp (5430N) and then it comes under Scottish Control. To find out the details, if interested, it is necessary to either consult the Aerad Supplement or the COM section of the UK Air Pilot.

From Pole Hill until reaching the next significant point in the flight, MARGO, the total distance is 49 nm. There are VOR/DMEs at each end of the stage POLE HILL to TALLA on B4 and from our previous assessment, they should both be within range for the whole stage. Navigation, therefore, should be quite simple. If fitted with twin VOR/DME, it will be possible to have a constant display of both TMG and Tr required as well as distance gone and distance to go. The DMEs fitted in many light aircraft, have a simple arrangement whereby, when locked on to a DME ahead of the aircraft, both the G/S and time to go can be displayed. In this case, Talla would be providing the means for a constant up-date of the ETA.

A point about this route to be noted is that the box surrounding the airway identifier or designator B4 is arrowed to indicate that this route is only normally available for traffic in the NWly direction. It is an interesting exercise to try to work out the route to follow for the reverse flight.

From MARGO, instead of flying along B4, the most direct path that could be followed would be the ATS route to Turnberry. The track is shown as 305°(M) and distance 78 nm.

If clearance was given for this route, how would it be navigated? Even if the aircraft only had the very basic navigation aid 'fit' required for airways flight, navigation of this route would be quite simple. The arrival at Margo would be identified by the readings on the Talla VOR/DME of 345°M/52 nm. Prior to this, assuming another VOR/DME available on the aircraft, TRN VOR should have been tuned

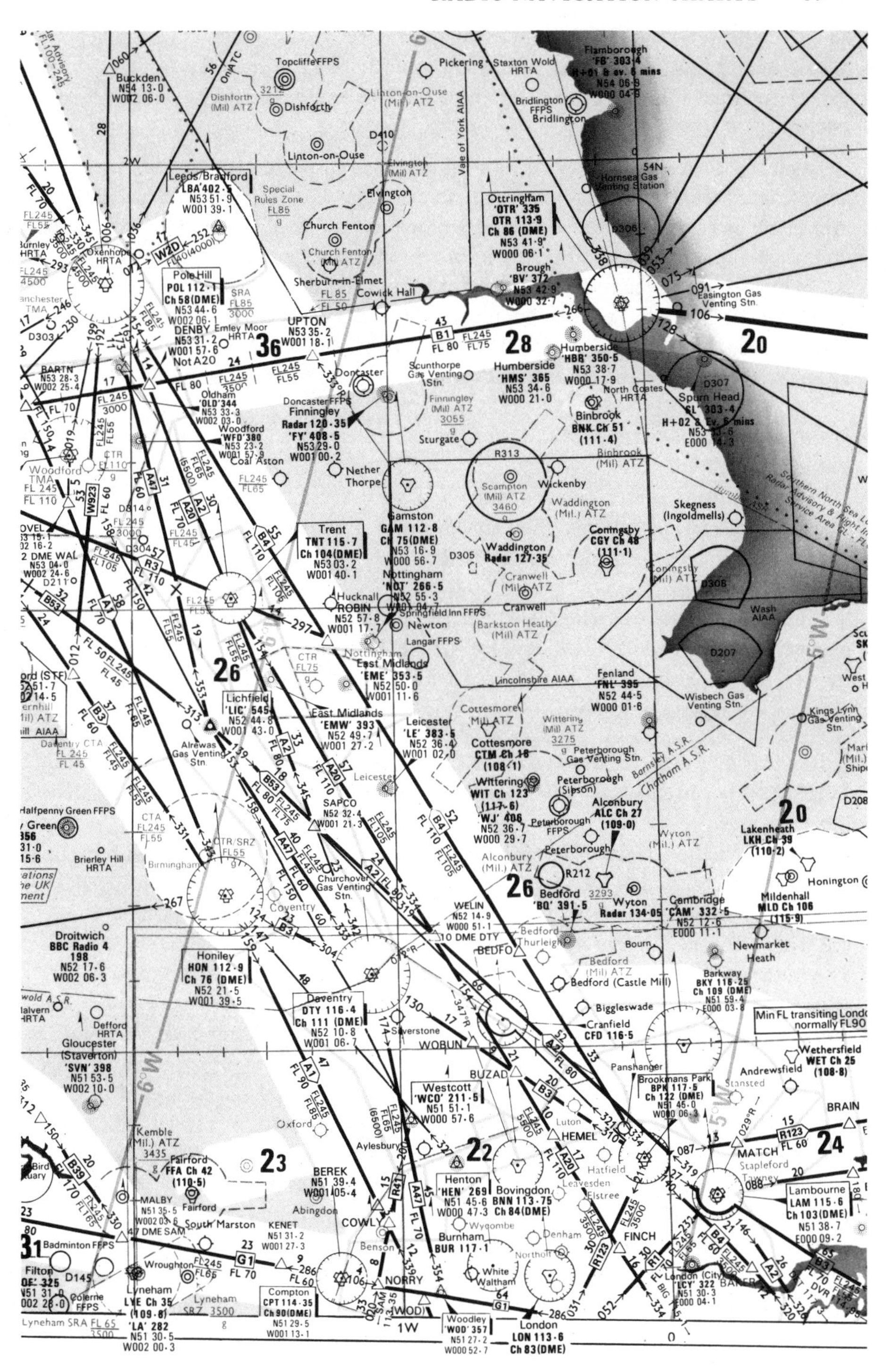

Topcliffe FFPS
Pickering
Staxton Wold HRTA
Flamborough 'FB' 303·4
Dishforth (Mil) ATZ
Dishforth
Linton-on-Ouse (Mil) ATZ
Bridlington FFPS
Bridlington
Linton-on-Ouse
Vale of York AIAA
Leeds/Bradford 'LBA' 402·5 N53 51·9 W001 39·1
Special Rules Zone
Elvington
Hornsea Gas Venting Station
Ottringham 'OTR' 335 OTR 113·9 Ch 86 (DME) N53 41·9 W000 06·1
Church Fenton
Church Fenton (Mil) ATZ
Brough 'BV' 372
Sherburn-in-Elmet
Cowick Hall
Pole Hill POL 112·1 Ch 58 (DME)
Easington Gas Venting Stn.
DENBY
Emley Moor HRTA
UPTON N53 35·2 W001 18·1
Humberside 'HMS' 365 N53 34·6 W000 21·0
Humberside 'HBB' 350·5 N53 38·7 W000 17·9
BARTN N53 28·3 W002 25·4
Oldham 'OLD' 344
Scunthorpe Gas Venting Stn.
Doncaster
North Coates HRTA
Spurn Head 'SL' 303·4
Woodford 'WFD' 380
Doncaster FFPS Finningley Radar 120·35 'FY' 408·5 N53 29·0 W001 00·2
Finningley (Mil) ATZ
Sturgate
Binbrook BNK Ch 51 (111·4)
Binbrook (Mil) ATZ
Coal Aston
Nether Thorpe
Scampton (Mil) ATZ
Wickenby
Waddington (Mil.) ATZ
Skegness (Ingoldmells)
Southern North Sea Low Level Radar Advisory & Flight Information Service Area
Gamston GAM 112·8 Ch 75 (DME)
Trent TNT 115·7 Ch 104 (DME) N53 03·2 W001 40·1
Waddington Radar 127·35
Coningsby CGY Ch 48 (111·1)
Nottingham 'NOT' 266·5
Cranwell (Mil.) ATZ
Cranwell
Hucknall ROBIN N52 57·8 W001 17·7
Springfield Inn FFPS
Newton
Langar FFPS
Barkston Heath (Mil) ATZ
Wash AIAA
Lincolnshire AIAA
East Midlands 'EME' 353·5 N52 50·0 W001 11·6
Fenland 'FNL' 395 N52 44·5 W000 01·6
Wisbech Gas Venting Stn.
Lichfield 'LIC' 545
Alrewas Gas Venting Stn.
East Midlands 'EMW' 393 N52 49·7 W001 27·2
Leicester 'LE' 383·5 N52 36·4 W001 02·0
Cottesmore (Mil) ATZ
Cottesmore CTM Ch 18 (108·1)
Wittering (Mil) ATZ
Peterborough Gas Venting Stn.
Kings Lynn Gas Venting Stn.
Barnsley A.S.R.
Chatham A.S.R.
Wittering WIT Ch 123 (117·6)
Peterborough (Sibson)
Halfpenny Green FFPS
SAPCO N52 32·4 W001 21·3
Leicester
'WJ' 406 N52 36·7 W000 29·7
Peterborough FFPS
Alconbury ALC Ch 27 (109·0)
Wyton (Mil.) ATZ
Lakenheath LKH Ch 39 (110·2)
Brierley Hill HRTA
Birmingham
Churchover Gas Venting Stn.
Alconbury (Mil.) ATZ
Peterborough
Bedford 'BO' 391·5
Wyton Radar 134·05
Cambridge 'CAM' 332·5 N52 12·6 E000 11·1
Honington
Mildenhall MLD Ch 106 (115·9)
Coventry
WELIN N52 14·9 W000 51·1
10 DME DTY
BEDFO
Bedford Thurleigh
Bourn
Newmarket Heath
Droitwich BBC Radio 4 198 N52 17·6 W002 06·3
Honiley HON 112·9 Ch 76 (DME) N52 21·5 W001 39·5
Bedford (Mil) ATZ
Bedford (Castle Mill)
Barkway BKY 116·25 Ch 109 (DME) N51 59·4 E000 03·8
Daventry DTY 116·4 Ch 111 (DME) N52 10·8 W001 06·7
Biggleswade
Cranfield CFD 116·5
Min FL transiting London normally FL90
Defford HRTA
Gloucester (Staverton) 'SVN' 398 N51 53·5 W002 10·0
Silverstone
WOBUN
Wethersfield WET Ch 25 (108·8)
Andrewsfield
Panshanger
Brookmans Park BPK 117·5 Ch 122 (DME)
Stansted
BUZAD
Westcott 'WCO' 211·5 N51 51·1 W000 57·6
Luton
HEMEL
BRAIN
Kemble (Mil.) ATZ
Oxford
Aylesbury
MATCH
Stapleford Tawney
Fairford FFA Ch 42 (110·5)
BEREK N51 39·4 W001 05·4
Henton 'HEN' 269
Bovingdon BNN 113·75 Ch 84 (DME)
Hatfield
Leavesden
Elstree
Lambourne LAM 115·6 Ch 103 (DME) N51 38·7 E000 09·2
MALBY
Abingdon
Wycombe
Denham
FINCH
Badminton FFPS
South Marston
KENET N51 31·2 W000 27·3
COWLY
Benson
Burnham BUR 117·1
Northolt
Wroughton FFPS
White Waltham
London (City) 'LCY' 322
Filton
Colerne FFPS
Lyneham LYE Ch 35 (109·8) 'LA' 282 N51 30·5 W002 00·3
Lyneham SRZ 3500
Compton CPT 114·35 Ch 90 (DME) N51 29·5 W001 13·1
NORRY
WOD
Woodley 'WOD' 357 N51 27·2 W000 52·7
London LON 113·6 Ch 83 (DME)
Lyneham SRA FL 65 3500

Fig. 6.6

in and identified and a Tr(M) to steer of 305° selected. On arrival at Margo, it would then simply be a matter of turning to make good the new Tr. In practice, the turn would probably have been started just before arriving at Margo so as to turn smoothly on to the new track.

A larger scale Area Chart is available for busy areas shown on these charts by black double-lined rectangles: London area is an example. Also, each airfield has its Approach charts. Area charts are simply blown up charts of the areas shown on the Aerads in a 'picture frame'. Further information is given to ease the pilot's load when in a busy area, calling for prompt attention to Control instructions. There is no change in symbols, etc., but holding points and patterns, communication frequencies for the various airports when departing therefrom or arriving thereto, special charts for specific aerordromes with their departure/arrival routes are clearly shown.

Approach charts simply blow up the facilities at a particular aerodrome, and give every detail about them.

High Altitude charts, labelled with an H, such as H108/109, are similar to the types we have been studying. They are for flight at or above FL 250, and since greater speeds are involved, a smaller scale is used so giving a larger ground area on each chart.

Tracks, distances, bearings, frequencies: all these change from time to time, changes incorporated every 28 days by AIRAC Notams. So, if any of the above information does not match your chart, do not worry, so long as you understand what we are trying to convey.

Before leaving the topic, it is convenient to run over Transition Altitude, Transition Level and Transition Layer. On take-off and landing, climb and descent, they are quite distinct.

The Transition Altitude (TA) is the altitude (QNH-based) in the vicinity of an aerodrome below which the vertical position of an aircraft is controlled using altitude (QNH-based). Transition Level is the lowest FL available for use which is physically (not necessarily numerically) above the TA. For example, taking-off at an aerodrome where QNH is 1 000 mb and TA is 4 000 ft, the pilot will use an altimeter setting of 1 000 mb when complying with ATC instructions regarding the altitudes at which to fly. On reaching altitude 4 000 ft, he will reset his altimeter sub-scale to 1 013 mb and so increase the indicated altitude by approximately 13 x 30 = 390 feet to 4 390 feet. It is likely that the first FL available will be 50 (5 000 ft on an altimeter set to 1 013 mb) and this will be the Transition Level. When at the end of a flight an aircraft approaching this aerodrome is cleared from an FL to an altitude during its descent, the current QNH will be given. On leaving its present FL, the pilot will change his altimeter setting to the QNH unless further FL passing reports are required. Once below the Transition Level altitudes will always be used. The Transition Level is the very latest level at which the change from the standard setting of 1 013 mb to QNH will be made.

The Aerad charts are of course constantly being brought up to date, and the operational use of the very latest issue is imperative.

Now try this exercise, using Fig. 6.8 and then any available EUR 2 for the latter part.

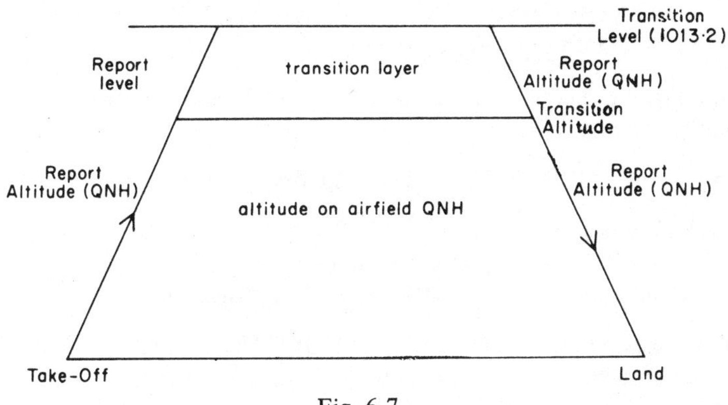

Fig. 6.7

You are planning a flight from STUTTGART (4842N 0913E) to REIMS (4919N 0403E) via R11, R7 and R10. You join the airway R11 at TANGO reporting point, and leave at MONTMEDY VOR. TAS 280 kt. You are equipped with twin ADF, VOR and DME. Descent from MONTMEDY VOR.

1. In absence of any other information, are you expected to fly ODD flight levels or EVEN flight levels?
2. Give a complete list of compulsory reporting points.
3. State TANGO NDB frequency, ident and type of emission.
4. Is there any facility at TANGO from which you could receive range information? If so, state how you would use it.
5. About half way between TANGO and ROTWE a grey line running N/S cuts your track. What purpose does this line serve?
6. Arriving at ROTWE reporting point how would you check arrival at ROTWE if TGO VOR/DME was not operating?
7. Between ROTWE and STRASBOURG:
   (a) what is your track?
   (b) what is the distance?
   (c) below airway centre line, this information is given:

$$\frac{\text{FL 240}}{6000}$$

give the meaning of each item.

18 nm from STRASBOURG you cross a broken black line and 10 nm later a black cross.

8. What is this black line for, and why is it broken?
9. What does the cross signify?
10. Just before crossing this line, in whose TMA are you flying?
11. Are you also flying in any CTR at this time?
12. How would you know when you are overhead STRASBOURG?

You leave STRASBOURG VOR at 1000 hrs on heading 322°(M) for GROS

TENQUIN. Your VOR is tuned to STR frequency with 315° on OBS your ADF is also tuned to STR. TAS say 280 kt.

13. Assuming you remain on track, what readings do you expect on the VOR and Relative Bearing Indicator?

You arrive overhead GROS TENQUIN at 1011 hrs.

14. What W/V was experienced on this leg?
15. What is your ETA at next reporting point?
16. How would your VORs and ADFs be tuned on this leg?

17. You are cleared to overhead LUXEMBOURG VOR, and arrive at 1022, FL 180. Give your full position report. (Assume the W/V remains unchanged.)
18. What information would you expect to receive in return?
19. When at MONTMEDY VOR who would you pass your position report to, and request permission to leave airway and proceed to REIMS? On what frequency?
20. Are you now in PARIS TMA?
21. On your last leg to REIMS while descending what is your minimum clearance altitude?
22. What navigation facilities are available at REIMS which you could use?
23. Between GROS TENQUIN and LUXEMBOURG the route is covered by some kind of restricted airspace. What can you tell about it?
24. Why is Germany shown with a white background whereas France, Belgium and Luxembourg have areas of grey?
25. Does any of the flight pass through BRUSSELS FIR? If so, where did you enter and leave?

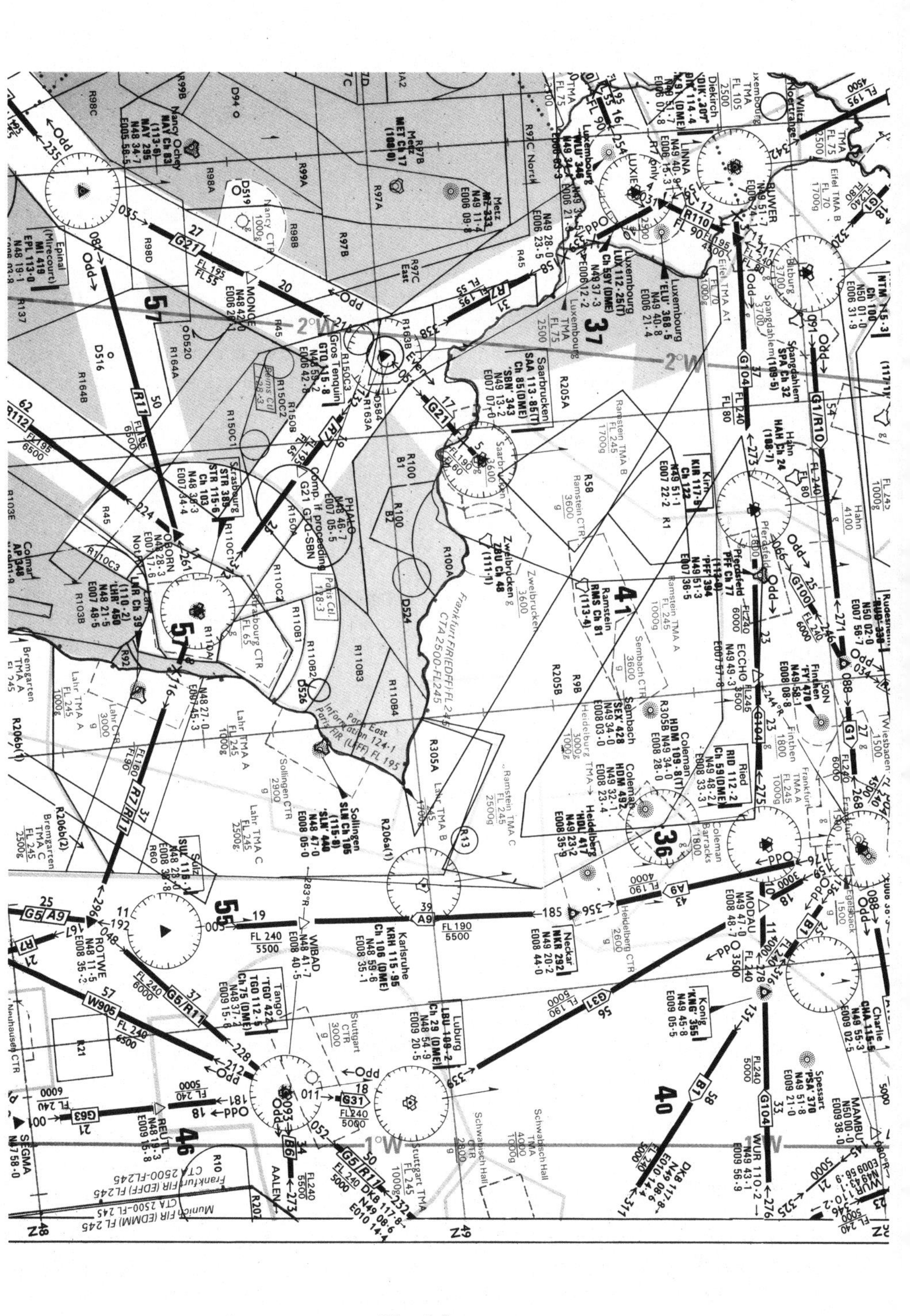
Luxembourg
Saarbrucken
Strasbourg
Karlsruhe
Ramstein
Sembach
Heidelberg
Coleman
Sollingen
Zweibrucken
Spangdahlem
Hahn
Frankfurt FIR (EDFF) FL 245
CTA 2500-FL245
Munich FIR (EDMM) FL 245

Fig. 6.8

7: Relative Motion

Relative Speed

Whenever two aircraft are in motion, each aircraft has a speed which is relative to the other. Two aircraft flying in formation at a given speed, have a relative speed with regard to each other: it is zero speed, because neither aircraft is going ahead or falling behind. If two aircraft are approaching on reciprocal tracks, aircraft A doing a speed of 150 kt and aircraft B speed of 200 kt, then each aircraft has a speed of 350 kt relative to the other. We call this the closing speed. Again, if an aircraft A with ground speed of 350 kt is overtaking aircraft B having a ground speed of 300 kt, each aircraft's relative speed is 50 kt, whereas aircraft A's closing speed is 50 kt. Therefore, if the two aircraft were initially separated by, say, 100 nm aircraft A will be alongside B in two hours.

Example. Two aircraft A and B are initially separated by 200 nm and are approaching each other. A's ground speed is 300 kt, B's ground speed is 260 kt. How long will they take to meet?

Relative (or closing) speed = 300 + 260
= 560 kt

Time taken to meet = 200 nm at 560 kt
= $21\frac{1}{2}$ min

How far would each have flown before they meet?

Aircraft A with ground speed of 300 kt will travel 107 nm in $21\frac{1}{2}$ min
Aircraft B with ground speed of 260 kt will travel 93 nm in $21\frac{1}{2}$ min

Practice problems

1. Two planes take off from stations 500 nm apart. They meet in 44 minutes. If the first plane travelled 3/5th of the total distance, find the ground speed of each.
Answer: 410 and 273 kt

2. An aircraft A, flying at ground speed of 390 kt is overtaking aircraft B, flying at ground speed of 310 kt. Two aircraft are initially 50 nm apart.
(a) In how many minutes will aircraft A be 5 nm behind B?
(b) When will aircraft A be 3 minutes behind B?
Answer: (a) $33\frac{1}{2}$ minutes; (b) $34\frac{1}{2}$ minutes

Relative Direction

When two aircraft are moving, each will have a relative direction with respect to the other. If two aircraft A and B left a point X at the same time, A on a track due North and B on a track due East, then to an observer in aircraft A, B will appear to be tracking SE. To an observer in aircraft B, A will appear to be tracking NW; this is A's relative direction with respect to aircraft B.

Relative Velocity

When a body has velocity, it has both speed and direction, and problems of relative velocity are essentially solved by accurate scale plotting. The first step is to bring to a stop the aircraft which wishes to observe the other.

In above example, say A's ground speed is 400 kt, B's ground speed is 320 kt. To find B's relative velocity with respect to A, bring A to a stop. This is done by imparting A's negative velocity to B. B, therefore, now has two velocities, its own velocity and A's negative velocity. Fig. 7.1.

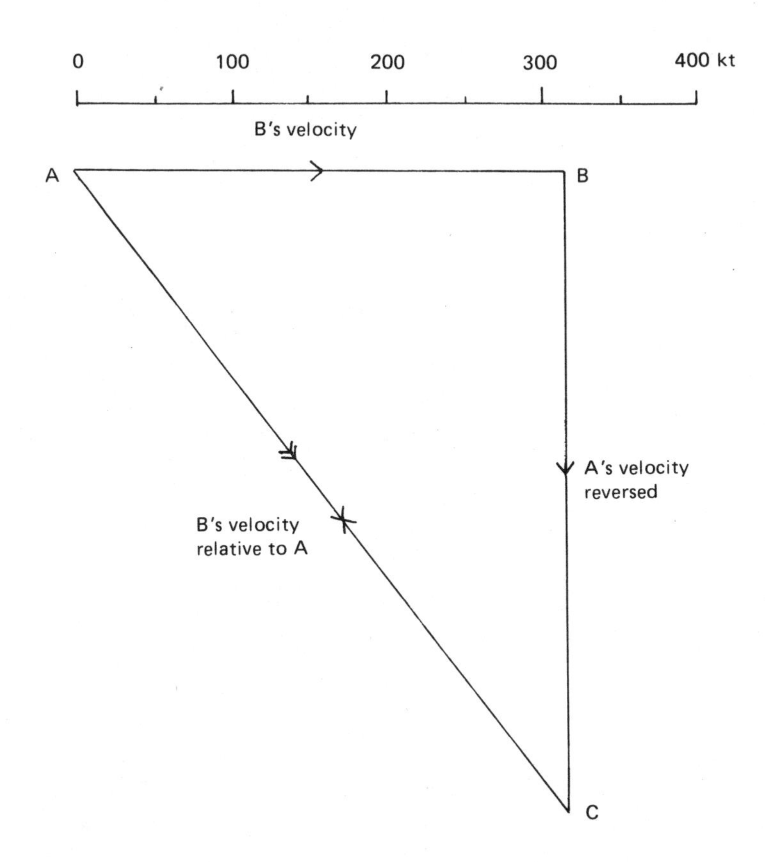

Fig. 7.1

In Fig. 7.1, AB is B's own velocity, that is, one hour's ground speed due East. BC is A's negative velocity, 400 kt due south. Complete the triangle ABC. AC is the relative velocity of B with respect to A which is 141° at 512 kt (by actual measurement).

If A's relative velocity with respect to B is required, B is brought to rest by imparting its negative velocity to A.

Where two flights do not originate at the same point as above, the relative velocity may still be estimated as in the following example (Fig. 7.2).

Given: Aircraft A's track is 050(T), G/S 210 kt. Aircraft B which bears 140(T) from A is doing a track of 355(T) at G/S250 kt. B is 200 nm from A. Find the relative velocity of B from A.

In these problems since the distance between two aircraft is generally very small in comparison with aircraft speeds, it is necessary to select two different scales, one for ground speed and one for distance in order to separate the two aircraft for neat

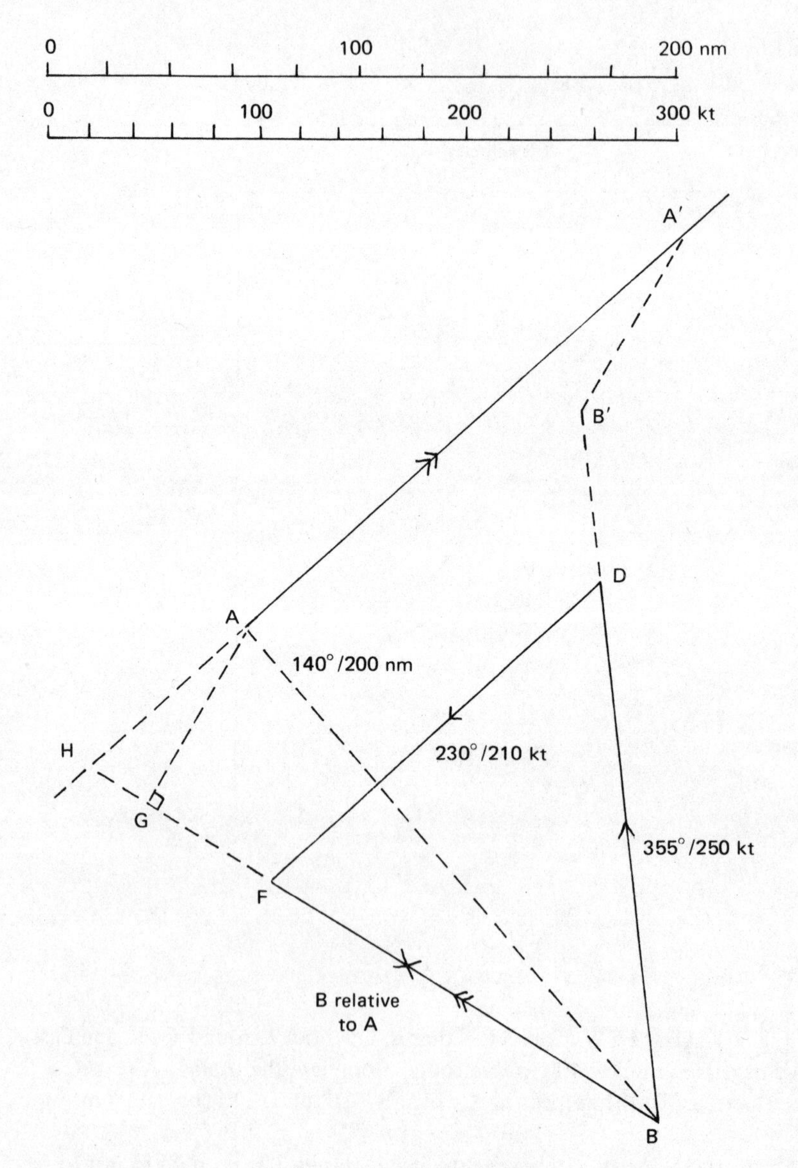

Fig. 7.2

plotting. Choose simple scales such as 1 cm = 50 kt for speeds, and 1 cm = 20 nm for distance.

Steps

1. Choose a convenient point A and plot from here A's Track.
2. From A, plot B on given bearing (140° in above illustration) at a given distance (200 nm) at distance scale – AB.
3. From B, plot B's velocity, BD.
4. From D draw in A's reversed velocity, DF.
5. Join BF

BF is B's relative velocity with respect to A – 302° at relative speed of 215 kt.

The above figure tells us more than just the relative velocity.

Produce BF to intersect A's track at H. If this line intersects A's track exactly through A's position, two aircraft are on such headings and speeds that a collision will occur. If BF produced crosses A's track behind A's position, as in above illustration, B will pass behind A. The shortest distance that A and B will be apart is indicated by AG drawn perpendicular from A to BH, measured at distance scale. Similarly, if BF produced crosses A's track ahead of it, B will pass ahead of A.

The above is true even if distance AB initially is not known.

The time that B will be shortest distance from A is calculated from known information of B's relative ground speed and the distance BG.

In Fig. 7.2, it is 190 nm at 215 kt = 53 min.

The actual positions of A and B after 53 min are shown on the diagram as A′ and B′. Note that A′B′ is equal and parallel to AG.

Interception

Interceptions are carried out on the principle of maintaining a constant relative bearing. Aircraft A sights another aircraft B on a relative bearing of 140°. If A wishes to intercept B, it must maintain such a heading and TAS that would result in B's relative bearing remaining constant throughout. Fig. 7.3.

In the figure, it will be noticed that if the bearing is not maintained constant, aircraft B is either going to fall ahead or behind of A.

In order to carry out an interception, the first step is to establish the positions of both aircraft and plot them. The line joining the two aircraft at any given instant is the bearing that must remain constant. In Fig. 7.4, this is the line AB, and is known as the Line of Constant Bearing.

To calculate the heading to intercept, draw in from A a vector AC to represent B's air velocity reversed. Centre C and radius A's TAS describe an arc to cut AB at D. ACD is the relative motion triangle and AD represents A's motion relative to B during the interception. From A draw in a line parallel to CD to cut B's heading line at E. E will be the air position at which the interception will occur and to find the Ground Position it would be necessary to apply the wind effect for the time to intercept. This can be calculated either by using distance AB and relative speed AD, or distance AE and A's TAS. Both should give the same answer!

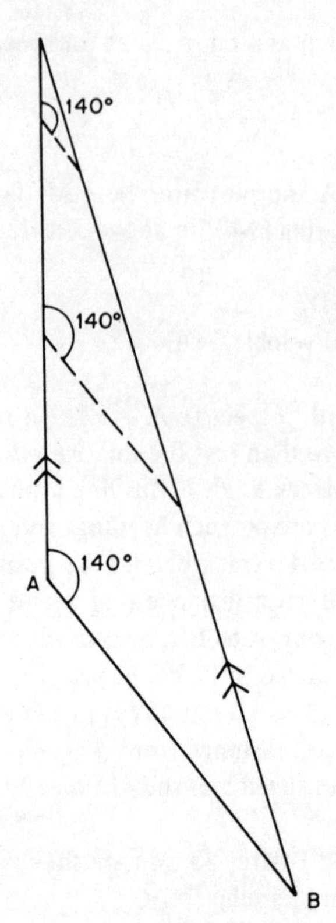

Fig. 7.3

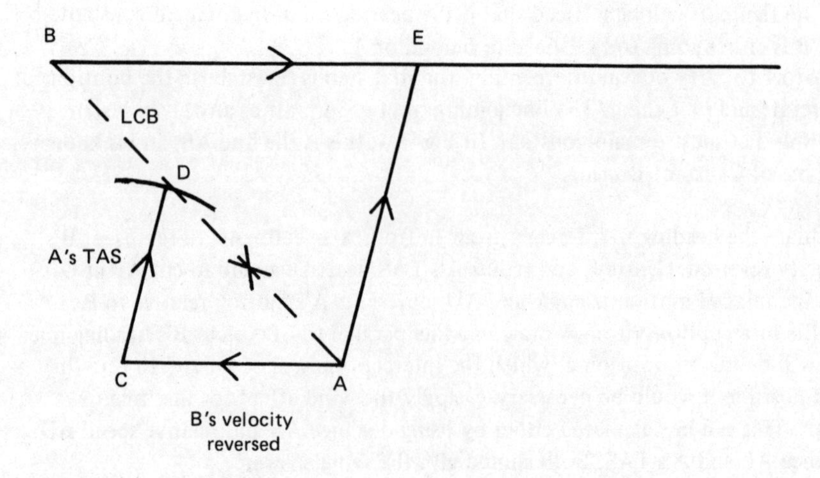

Fig. 7.4

Latest time to divert to an alternate (Point of No Alternate – PNA)
To find the latest time to divert to an alternate would be worked out on these lines. An aircraft with 5 hours endurance has an alternate to beam of Track. Join departure point A to the alternate B, measure the distance and derive its hypothetical G/S from the endurance. Along both Tracks, using the same vector scale for both, plot the distance covered in one hour; this is the Line of Constant Bearing GF. Produce GF. Still with same vector scale, at A, plot a W/V vector and from it with radius TAS, strike an arc on GF produced at K. Join AK. AK paralleled from B to cut the outbound Track is the Ground Position of Turning, and the plot contains all the necessary information for the diversion.

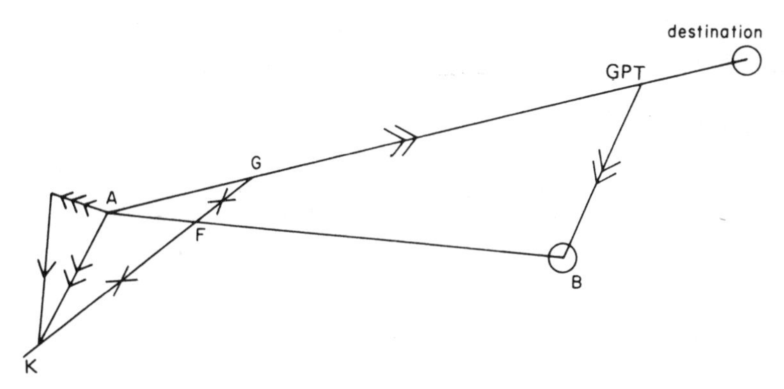

Fig. 7.5

Problems

1. An aircraft A in position 55N 02W is making good a track of 180(T) at ground speed of 210 kt. Aircraft B is flying along parallel of 49°10′N and at 1000 hrs, it bears 220° from A. Both aircraft eventually meet.
   What is B's bearing from A at 1030 hrs?

Answer: 220°(T) (Bearing must remain constant)

2. Aircraft A making good a Tr 030(T) at 300 kt passes over B which is making good a track of 310 at 260 kt. Estimate relative velocity of B from A prior to passing over B.

Answer: 255° – 361 kt

3. Aircraft A on track of 090 is doing ground speed of 200 kt. It will collide with B in 5 min if no alteration to heading is made. B at present is on bearing of 030 relative from A. If B's ground speed is 250 kt, estimate B's track and relative velocity, assuming zero drift.

Answer: Tr 323; Rel.Vel. 300° – 402 kt

4. Aircraft A whose TAS is 270 observes aircraft B on relative bearing of 040°. Aircraft B observes A on relative bearing of 330°. If two aircraft are on collision headings, determine
(a) TAS of aircraft B;
(b) speed at which the two aircraft are closing.

Answer: (a) 347 kt (b) 507 kt

5. An aircraft A is flying due south at ground speed of 170. B is flying East at ground speed of 210 both having left the same point. What is the relative velocity of B with respect to A?

Answer: 051° – 270 kt.

6. Aircraft A is heading 030(T), TAS 230. Aircraft B which bears 310(T) from A is heading 080(T) at TAS 320 kt. If neither aircraft alters heading or speed, which will pass ahead of the other? What is B's motion relative to A?

Answer: Aircraft B. 305°/246 kt.

Section 3

FLIGHT PLANNING

8: Principles of Flight Planning

Introduction

Before flight in the commercial business of carrying passengers or cargo for hire or reward, a very comprehensive Flight Plan must be made, giving the Headings Magnetic to steer, the time on each leg, the fuel to be consumed, the height to fly, the alternates available, and any other detail useful for the trip. It is a plan, a guide, and its main purpose is safety, ensuring primarily that sufficient fuel is uplifted plus a bit extra for mother. In the air, amendments to Headings and times will be made, with a continuous check on fuel consumption and weather ahead, by actual navigation.

The first prerequisite on arrival at the field is to obtain the latest Met information for the route, and for all the aerodromes likely to be used; forecast Wind Velocities and temperatures at pertinent heights will be given, and from these, the Flight Plan can be filled in. This done, adequate fuel can be ordered and other matters such as range, point of no return can be duly entered. The complete plan will be reported to Air Traffic Control, so that in the air a full surveillance of the aircraft's progress will be kept.

A couple of lines of a Flight Plan might look like the example opposite.

Having gaped at that lot for a moment or two (and perhaps checked the TAS, times, Headings on your computer), you will appreciate that here is most of the information required for the trip and on the trip; but of course, temperatures, Wind Velocities are forecast, fuel consumption may not go according to the book, ETAs will invariably change, but the Plan is there.

You will have remarked that fuel consumption has decreased slightly on the second leg: as the weight of the aircraft decreases with fuel being burned off, so there is less weight to heave through the air, and consumption will be reduced. It is commonsense that as an aircraft gets lighter, with the fuel being consumed, its performance will improve assuming that there is no significant wind change. By an improved performance, we mean a smaller value for the ratio of fuel flow to TAS. This is usually expressed as nm/kg, often referred to as the economy figure. Modern turbo-jet aircraft are usually operated at a constant Mach number – in fact on some routes it is an ATC requirement that this is done. Usually the Captain has very little choice in the speeds at which he may operate. As set out in the Operations Manual for a modern tri-jet it is:

(i) *High Speed (HS) Cruise* Flying at Mach 0.85 subject to not exceeding the engine limitations.
(ii) *Long Range (LR) Cruise* Flying at Mach 0.82

In both cases, of course, the economy will improve as the aircraft gets lighter.

Apart from the speed selected and the wind, over which we have no control,

STAGE		Press height ft × 1000	RAS kt	Temp °C	TAS kt	W/V	Tr (T)	Hdg (T)	Var	Hdg (M)	Dist nm	G/S kt	Time min	ETA	Fuel flow kg/hr	Fuel kg
From	To															
ALICE SPRINGS	Abm OODNA	31	264	−45	423	270/60	149	156	5E	151	237	450	32	0447	3900	2080
Abm OODNA	LEIGH	31	264	−46	423	260/60	149	156	6E	150	231	442	31	0518	3860	2000

Note On Computers such as Airtour CRP 5, when the TAS calculated from RAS using Pressure Altitude and Temperature exceeds 300 kt, a further correction has to be made using the window on the slide rule marked comp. corr. (compressibility correction) – see handbook for the computer. Neglecting this correction would give a TAS 10 kt too high in this case.

the other factor having a marked effect on the economy is FL. Generally speaking with modern turbo-jets it is best to fly as high as is possible and permitted. This should be done even to the extent of increasing level as the aircraft gets lighter. Maximum economy would be achieved by a gradual upward drift throughout cruising flight. This cruise–climb technique is used by supersonic aircraft but in the more crowded levels, ATC prefer constant level cruising.

It is possible, that at intervals, an aircraft may be permitted to 'step up' to the next available FL (usually 4 000 ft higher). This is only likely every 3 to 4 hours even with the biggest modern aircraft.

It is quite often stated that air temperature affects operating economy. This is not, in general, of practical significance. Cruising at constant Mach number, a higher temperature produces a higher TAS at the expense of more engine power and so higher fuel flows. Practical tests show that the two effects tend to balance out producing no significant change in the overall economy (kg/nm). The higher TAS will, of course, give a slightly better flight time.

The Principles of Flight Planning

In the very first place, with the information available before flight, the problem is one of work with the computer to resolve this information into the essentials for the flight itself. After this, as the pilot considers fuel, it is aircraft weight which is the governing factor; this weight decreases dramatically as the trip progresses with the high rate of fuel consumption, and a mean consumption or a mean TAS *per sector* must be deduced. Aircraft manufacturers set out this data in the aircraft manual, and for a 'weight at start' of a leg will, for the altitude and temperature, proffer a consumption for the next hour, or even for only the next half-hour, as we shall see. Such data sheets are used in the ATPL exam, but in CPL the knowledge of principles is introduced, and the averages of consumption or TAS must be calculated.

Let us consider first a Flight Plan for a voyage where the TAS is reasonably constant, or in other words, is given, but the fuel consumption varies as the aircraft's weight reduces.

A snippet from the data:

Consumption (kg/hr) at varying weights (kg)			
70 000	65 000	60 000	55 000
4 200	4 000	3 900	3 800

Take-off wt: 68 500 kg

Climb: Mean TAS 340 kt, time 38 min, fuel used 3 900 kg

The body of the flight plan – RAS, G/S, times, distance covered on the climb and descent, ETAs, can be completed straight off. The weight at start is 68 500 kg, so at the top of climb, having used 3 900 kg, it will be <u>64 600</u> kg for the commencement of level flight. We need the mean weight on the next leg in order to estimate as closely as possible the average consumption on it. Assume for ease of example that the leg will take one hour; from the table above, after half an hour, about 2 000 kg will have been burnt off, giving a weight in mid-leg of 62 600 kg; using this figure to enter the table, a mean consumption will be extracted of 3 950 kg, rounded off to avoid pedantic digits.

To proceed, the weight at start of the next leg is 60 650 kg (having flown for one hour at a consumption of 3 950 kg/hr from a start of 64 600 kg): say the leg will take 40 min. After 20 min at 3 910 kg/hr – and do not strain here, visual and mental calculation is enough – 1 300 kg will be burnt off and the mean weight may be taken to be 59 350 kg. This weight from the tables gives a consumption of 3 890 kg/hr, which is entered on the plan, and in 40 min will use 2 590 kg, giving a weight at start of the next sector of 58 060 kg. And so on.

A Flight Plan with a constant consumption but varying TAS is somewhat more involved; the plan must be done line by line, and the estimation of weights to find a mean TAS has its hazards, since no times on the legs are available.

A snippet from the data:

Consumption	Mean Weight (kg) v. TAS (kt)			
kg/hr	79 000	77 000	74 000	72 000
2 400	300	308	319	326

Weight at commencement of climb: 81 000 kg
Climb: Mean TAS 235 kt, time 48 min, fuel used 2 250 kg

The climb leg can be solved for G/S, time, distance and the rest, and the weight at start of level flight is 78 750 kg.

The best technique is to commence by estimating an approximate economy (kg/nm) figure. In this case, using conveniently rounded off figures of 2 400 kg/h and 300 kt, the economy figure would be 8 kg/nm in still air. If the next stage was 190 nm, the appropriate fuel used for *half* the stage would be 4 x 190 = 760 kg. The mean weight would, therefore, be 78 750 – 760 = approx 78 000 kg.

Use this weight to extract the TAS from the table = 304 kt and complete the leg of the Flight Plan. The actual fuel used will be 1 560 kg. Check. The weight at start of next leg is 77 190 kg, whence to continue the exercise.

On finishing the final line to destination, the fuel used from departure to destination can be totted up (Item A), and the weight over the destination field calculated; this weight, less any fuel used for final descent and landing, will give the anticipated landing weight.

In calculating the fuel or TAS for the alternate from the data given, it is sufficient to use the weight at destination as a leader into the requirements for the leg.

The fuel to destination plus alternate fuel plus contingency fuel plus taxi, takeoff, circuit, landing plus any other reserves or percentages will give the total fuel required for the trip; from this figure, the payload can be worked out, and this is what keeps us in business.

All this, the type of Flight Plan in qualifying exams for pilot licences, may have astounded you in its vagaries and guesswork; have no fear, it is but a presentation of the principles of the stuff to the student pilot: the manuals are of course much more precise, full of information garnered from tests and checks carried out with care and accuracy, but still on mean weights as shown, though for a specified period of time. We will move on to this practical matter at once.

Presentation of Data

Every aircraft type produces a cruise control manual, wherein all information at all heights at all temperatures for each specific purpose is shown, either graphically or tabulated, the latter by far the more popular. Climb, short range diversion, level cruise by appropriate methods, 4-engines, 3-engines, 2-engines, level cruise; in fact anything that the pilot requires for his aircraft for his route, presented succinctly, for rapid production of the Flight Plan with the station manager breathing down his neck to get him away, the fuel wallah palpitating for the fuel requirements, the load people agitating for pay load particulars. You won't get the aircraft type on your licence till you've mastered the manual, but for examination purposes, the Civil Aviation Authority has produced some Data Sheets set out along the accepted lines. Data Sheets 33 are part of your equipment, so we'll refer to them constantly and work out a Flight Plan sample.

The performance of an aircraft is dependent on pressure and temperature which in turn determine air density. The pressure is conveniently expressed as the pressure altitude, i.e. the altimeter reading with 1 013 mb on the sub-scale. The temperature is normally described by the temperature deviation which is:

correct outside air temperature (COAT) – standard temperature

The standard temperature is the standard used for each particular FL in the preparation of the tables. In Data Sheets 33, these standard temperatures are detailed in Table 33F. This table differs from that used in most tables which generally use an approximation to the International Standard Atmosphere (ISA). The first step in working out a Flight Plan is to deduce the temperature deviation for each stage. When extracting performance for any part of the flight, it is important to check that the table being used is for the appropriate temperature deviation range as shown at the top of the table.

Let us now plough gently through the following Flight Plan, assisted by Data Sheets 33: remember the penalties are heavy for serious arithmetical inaccuracies in the exam as well as in the air: – you would feel a real charley to find in mid trip that you'd uplifted 1 000 kg too little fuel.

Information is as follows:

A flight is to be made from ROME TO ACCRA. LAGOS is the terminal alternate. Route details are given on the pro forma.

Loading:	Weight at start of take-off is 130 000 kg.
Climb:	Climb on track from 1 000 ft over ROME to flight level 340. (Table 33A).
Cruise:	Cruise at the levels given in the Flight Plan (Table 33C).
Descent:	Descend on Track to arrive over ACCRA at 1 000 ft. (Table 33E).
Alternate:	Use Table 33G. Assume diversion is commenced 1 000 ft over

FLIGHT PLAN

STAGE		Flight Level	Temp. Dev.	WIND		Track	Drift	Heading	T.A.S.	Wind comp.	G/S	Distance	Time	Fuel flow	Wt. at start	Fuel required
From	To		°C	Direction	Speed kt.	°(T)		°(T)	kt.	kt.	kt.	n.m.	min.	kg./hr.	Kg.	Kg.
TAKE OFF FUEL		←										→		↔		
ROME	Top of climb	↗	−3	040	20	170						244		—		
Top of climb	PALERMO	340	+2	340	50	170										
PALERMO	IDRIS	350	+4	290	70	180						331				
IDRIS	GHAT	350	+4	310	50	198						490				
GHAT	NIAMY	350	+8	280	80	213						824				
NIAMY	Top of descent	350	+12	210	15	197						487				
Top of descent	ACCRA	↘	—	200	10	197								—		
ACCRA	LAGOS	↗ ↘	—	200	20	075			—		—	217		—		

FLIGHT PLAN

STAGE From	STAGE To	Flight Level	Temp. Dev. °C	WIND Direction	WIND Speed kt.	Track °(T)	Drift	Heading °(T)	T.A.S. kt.	Wind comp. kt.	G/S kt.	Distance n.m.	Time min.	Fuel flow kg./hr.	Wt. at start 1000 kg	Fuel required kg
TAKE OFF FUEL		←										→	02	↔	130.0	1000
ROME	Top of climb	↗	−3	040	20	170	2S	168	379	+14	393	131	20	—	129.0	4400
Top of climb	PALERMO	340*	+2	340	50	170	1P	171	488	+49	537	244 (total); 113	12½	7050	124.6	1470
PALERMO	IDRIS	*350	+4	290	70	180	8P	188	486	+20	506	331	39½	6900	123.1	4540 (50* above)
IDRIS	GHAT	350	+4	310	50	198	6P	204	486	+17	503	490	58½	6710	118.5	6550
GHAT	NIAMY	350	+8	280	80	213	9P	222	486	−36	450	824	110	6430	112.0	11790
NIAMY	Top of descent	350	+12	210	15	197	0	197	494	−15	479	394	49½	6115	100.2	5050
Top of descent	ACCRA	↘	—	200	10	197	0	197	369	−10	359	487 (total); 93	15½	—	95.1	700
															Item A	35550
ACCRA	LAGOS	↗↘	—	200	20	075	2P	077	—	+12	—	217	35	—	94.4	3730

*Step.

ACCRA and ends at 1 000 ft over LAGOS.

Fuel: (i) Sufficient for take-off and climb to 1 000 ft over ROME, plus:
(ii) Sufficient for flight from ROME to ACCRA and to alternate LAGOS, plus:
(iii) 800 kg for circuit and landing, plus:
(iv) 9 000 kg reserve
(a) Complete the Flight Plan.
(b) What weight of fuel is required?

Before starting, note that computer work is reduced by Drift and Wind Component Tables, pages 24 and 25 of the Data Sheet leaflet. The TAS is regarded broadly to give sufficient accuracy for flight planning purposes: a set is provided for each aircraft, with its mean cruising speed, mean climb and mean diversion speed. Here we have two tables, 480 kt and 380 kt; the wind speed is set out across the top, with the angle between wind direction and Track down the side. Thus, a drift and wind component can be read off, though the port or starboard bit must be determined. Thus, Track 180(T), W/V 290/70, your expected TAS 486 kt – angle is 110° down the side, against wind speed 70, drift is 8, component +20; use G/S 506 kt, and with a southerly Track with rough westerly wind, drift is clearly port. The appropriate table can be used to press on with the Flight Plan.

Aircraft Weights

It is necessary to keep a running record of the aircraft weight. We will keep the record in thousands of kg (tonnes) to an accuracy of 1 decimal place (i.e. 100 kg). This will be quite accurate enough for entering the performance tables. Fuel amounts will be calculated to the nearest 10 kg. We start our weight record by putting the take-off weight, 130.0 tonnes, on the first line in the weight column. Turn to table 33A on page 4 (temp devn –5 to –1°C). Footnote 2 gives us our first line entries – time and fuel to 1 000 ft, 2 min and 1 000 kg. Subtracting 1 tonne from 130.0 gives the weight at the start of the main climb, 129.0, and this is entered on the ROME to TOC line in the weight column.

Climb

Table 33A is now entered with the take-off weight and the FL 340 for the first cruise stage. This is a surprising FL because under the semi-circular rules 340 is not generally used. Against the figure for 34 000 TAS 379 kt is read being the mean TAS for the climb, and along the same line under 130 000 kg, read off the fuel required and the minutes to reach 34 000 ft – 4 400 kg and 20 min. Write these in the plan and subtract 4.4 tonne from 129 to get the weight, 124.6, for the start of cruise. Using your calculator obtain the distance gone on the climb (20 min at 379 kt), 131 nm, and subtracting from the total stage length to PALERMO from ROME obtain the balance of the distance to go in cruising flight.

Level

All the time the temp dev and FL must be watched: there is absolutely no reason why one shouldn't move from one page to another. And as an obiter dictum; Flight Level is the same thing as Pressure Height. Here we go to Table 33C, Standard 0 to +9°C. The top line is separated into individual hours of cruise, and the side has again pressure height and mean TAS: if you started at 137 000 kg, at 34 000 ft and flew at that pressure height, provided the temperature did not go outside the limits for the table, one could go steadily along the line. In this case, our weight at start is 124 600 kg, at 34 000 ft, TAS is straight 488 kt, but we must interpolate for fuel flow between the column:

129 : 7 300 and 122 : 6 900

The columnar weight difference is 7 000 kg for a consumption difference of 400 kg

So: 124 600 – 122 000 = 2 600

$\frac{2\,600}{7\,000} \times 400$ gives 150 kg to the nearest 10 kg

∴ consumption for 124 600 kg initial weight is 6 900 + 150 = 7 050 kg

which enter, and complete the PALERMO line, and be careful where you enter the fuel required of 1 470 kg (that's what you made it, I hope). The biggest boobs in Flight Planning are invariably arithmetical, cocking up a thousand with a hundred digit.

Proceed now with a start weight of 123 100 kg to IDRIS, checking the temp dev, O.K., keep the same page: but the pressure height is now 35 000 ft. From the notes on page 2 of the Data Sheets, an en route climb of 4 000 ft is ignored for time, but add 200 kg to fuel used: to be perfect then, we need to chuck in 50 kg to the fuel required on the IDRIS leg for a 1 000 ft climb. From Table 33C, TAS 486 kt, and 6 900 kg is accurate enough for 123 100 kg weight at start. Complete the line, and you'll find 39½ min gives you 4 550 kg required, +50 kg, a round 4 600. Weight at start for GHAT 118 500 kg. Into the Tables again, check the temp dev, O.K. Interpolate as before, between 123 : 6 900 kg and 116 : 6 600 kg for a consumption at 118 500 kg aircraft weight.

$$\frac{2\,500}{7\,000} \times 300 = \frac{750}{7} = 110 \text{ kg}$$

to be added to the 116 000 weight consumption = 6 710 kg, and the TAS is still 486 kt.

Complete the GHAT and NIAMY lines. We now must deal with the descent line to find time and distance covered before we can find out how far along the Track NIAMY – ACCRA to fly before commencing the descent. This is Table 33E, and is as plain as a pikestaff: as we're leaving 35 000 ft, TAS is 369 kt, fuel used 700 kg, time 15½ min. Complete the descent line, enter distances for the level and descent bits, and now to finish off NIAMY – TOD: the temp dev is +12°C, so with a weight at start of 100 190 kg on the table marked 'Standard +10°C to +14°C', enter at 35 000 ft, TAS 494 kt, interpolate for 100 000 kg between 103 : 6 200 and 96 : 6 000, giving 115 kg to add to 6 000, giving 6 115 kg consumption. Finish off so far: add the fuel requirements from departure to destination, usually known as Item A, and above all check that your weight at ACCRA + this figure = 130 000 kg.

The alternate must now be dealt with, Table 33G. The explanatory notes are reasonably clear: enter the tables for your conditions, and then make corrections. We're at 1 000 ft, will climb to 35 000 ft and descend to arrive over LAGOS at 1 000 ft. From drift and wind component table for TAS 380 kt, extract drift 2P, wind component +12, enter 33G and with a spot of visual interpolation: 35 min, 3 620 kg. Corrections are: none for height, weight is 94 400 kg ∴ add 3%, 108 kg say 110, used 3 730 kg. The corrections are all straightforward, and they are set out for you: no need to learn them by heart; it is automatic to check them, though, in every Flight Plan.

All that needs be done now is to tot up the Fuel on Board requirements, as demanded by the question, or by the Station Officer on the route. It is wise to set it out as on a fuel chit.

Item A	35 560
Alternate	3 730
Circuit & Landing	800
Reserve	9 000
FOB	49 090 kg

This makes it easy to deduce the landing weight at ACCRA, for example: you will use the 800 kg for circuit and landing, but still have alternate and reserve fuel in the tanks, so landing weight = 130 000 – 36 360 = 93 640 kt.

Divers problems in Flight Planning (Data Sheets 33)
The quite practical type of problem that involves a trip of a certain distance, part at low altitude (29 000 ft or below), part at high, hold and descent, is straightforward once you know your way around the Data Sheets. The information presented to you in an examination or in the Briefing Room must be complete enough for an answer to be arrived at, and there should be no difficulty, for example, in working out a descent before solving the time and fuel for cruise: there's nothing new in that. Table 33E for descent couldn't really be easier. I'm not trying to offend your intelligence in reminding you that a hold is a hold, where a ground speed is unnecessary: in the artificial atmosphere of the exam room, it is easy to start hunting for the absurd like 'how far have I gone on the hold leg'.

Table 33D gives the Low Level Cruise information, and do, oh do, notice the footnote about the mean weight of 100 000 kg. A mean weight of 135 000 kg increases the consumption at 15 000 ft at ISA +7, by 705 kg/hr. The table otherwise is self explanatory, calling for only the simplest interpolation.

A climb from 15 000 ft to 34 000 ft in Table 33A demands a simple subtraction of fuel at 15 000 ft for the weight at start of climb from the fuel at 34 000 ft, but a visual interpolation is required at bottom and top for intermediate weights. Keeping ISA +7, with a weight of 133 840 kg at 15 000 ft to start the climb to 34 000 ft, the table says –

		135 000 kg		130 000 kg	
Press Height	Mean TAS	Fuel kg	Time min	Fuel kg	Time min
34 000	387	5 800	31	5 300	28
15 000	324	2 100	9	2 000	9

Fuel at 15 000 ft for 133 840 kg aircraft weight is 2 080 kg
Fuel at 34 000 ft for 133 840 kg aircraft weight is 5 700 kg
Fuel used for this climb, then, is 3 620 kg, and such a round-off figure is quite adequate, as is the similarly subtractive time of 22 min.

The mean TAS for this climb demands an entry into the graph labelled for the exercise as Table 33B: enter with top of climb height across to the appropriate start of climb height curve, drop to the Reference line, and then parallel up or down as far as the temp dev axis, and read off the mean TAS. Our example above gives 418 kt. Interpolation of the bottom of climb curve is visually done. Don't make a large theoretical chore of any of this: the table, anyway, gives an example, which is worth a moment's study.

Descent is plain sailing (Table 33E) just read off TAS, fuel used and time taken from altitude to 1 000 ft: if the bottom of descent is not 1 000 ft, subtract one fuel from t'other, ditto time; add the two TAS and subtract 290. Hold is taken at the altitude on Table 33D, and the fuel calculated for the time of hold.

Try this, using Data Sheets No. 33

An aircraft is to fly from A to B, a distance of 950 nm on a Track of 250(T). Take-off weight is 140 000 kg, and the aircraft will successively:

(i) Climb from 1 000 ft to 15 000 ft, and then cruise at this level for 25 min (Table 33D).
(ii) Climb from 15 000 ft to 34 000 ft and cruise at this level until a descent is made to arrive over B at 6 000 ft.
(iii) Hold at 6 000 ft over B for 30 min (Table 33D) and then descend over B to 1 000 ft.

Details of these stages, temp dev and W/V are given below.
Complete the Flight Plan, giving the total fuel required and the total time.

STAGE Press Alt	Temp Dev	W/V	TAS	Wind Component	G/S	Dist	Time min	Fuel Flow kg/hr	Start Weight kg	Fuel kg
T/O & climb to 1 000 ft	–	–	–	–	–	–		–		
1 000 – 15 000 ft	+3	200/40						–		
Level 15 000 ft	+6	240/50					25			
15 000 – 34 000 ft	+6	260/60						–		
Level 34 000 ft	+8	290/70								
34 000 – 6 000 ft	–	230/50						–		
Hold at 6 000 ft	+6	–		–	–	–	30			
6 000 – 1 000 ft	–	–		–	–	–		–		
						950				

Answer: 24 000 kg; 3 h 02 m (± 200 kg ± 2 min)

The following problem is really an exercise in figure manipulation and logical method, but it has a very practical role, for often the weight of the aircraft at destination is the limiting factor.

Using Data Sheets No. 33, a flight is to be made from A to B, distance 1 150 nm, to arrive over B at 6 000 ft at weight 98 000 kg

Climb On Track from 1 000 ft over A to FL 340 (Temp dev +6, head wind component 30 kt)

<u>Cruise</u> Four-engine level cruise at 0.86 Mach at FL 340 (Temp dev +6, head wind component 55 kt)

<u>Descent</u> On Track to arrive over B at 6 000 ft (head wind component 25 kt)

Give the time and fuel required for:

(i) Climb from 1 000 ft
(ii) Cruise
(iii) Descent

<u>Descent</u> first, from Table 33E:
TAS (367 + 300) – 290 = 377 kt:
∴ G/S is 352 kt. Time given 12 min, so distance 70 nm:
Fuel given 600 kg:
Weight at start of descent is therefore <u>98 600</u> kg.
<u>Level</u> next, Table 33C, temp dev +6°C:
TAS 488 kt ∴ G/S 433 kt.
The aircraft's weight is going to finish the cruise sector at 98 600 kg; in the Table, from 102 000 to 98 600 kg gives:
Fuel 3 400 kg.
At noted consumption of 6 200 kg/hr, this takes 33 min:
33 min at G/S 433 kt gives distance 238 nm
The preceding hour uses 6 400 kg and distance 433 nm
A mental check indicates that there may be little cruise distance left, so take a summary:
<u>After</u> the climb <u>and</u> an undetermined period of cruise, the all-up-weight is:
98 000 + 600 + 3 400 + 6 400 = 108 400 kg
Similarly, the distance <u>gone</u> is:
1 150 – (70 + 238 + 433) = 409 nm
To enter the climb table, the TOW is required, so this must at this stage be estimated as accurately as possible. A glance at the appropriate page of Table 33A suggests a 22 min climb at a TOW of 115 000 kg, fuel used 4 300 kg; the climb G/S 357 kt for this time means 131 nm will be covered on the climb, and 278 nm is left for the very first cruise bit. Continuing with this procedure, enter the level cruise Table, read off the consumption 6 700 kg/hr, calculate the time to do 278 nm at G/S 433 kt; thus, 38½ min and fuel used 4 300 kg. The approximate TOW is:
108 400 + 4 300 + 4 300 = 117 000 kg,
and although the climb table gives fuel and time from 1 000 ft which is just what the question demands, the top line is classified as TOW and 1 000 kg must be included in the TOW figure for the initial take-off climb. Entering the table then with 118 000 kg:

Climb takes 23 min, uses 4 500 kg, distance 137 nm. Level flight starts at 112 500 kg all-up-weight (the initial climb fuel of 1 000 kg being allowed for), and so this portion will take 38 min to fly the 272 nm at G/S 433 kt, at fuel consumption 6 600 kg/hr = 4 200 kg. There is an element of meaning-off the extracted figures from the entered figures in the tables for intermediate weights, but there is no need for pedantic precision.

The answers are:

(i) Climb: 23 min, fuel 4 500 kg
(ii) Cruise: 2h 11m, fuel 14 000 kg
(iii) Descent: 12 min, fuel 600 kg

Diversion and Hold

Another practical problem in this paper is a diversion arranged for you somewhere in the closing stages of a trip. As the examiner wants to know if you are really familiar with the tables, he will divert you half way down the descent, and give you a hold. At one fell swoop, he's got you in every table in the book, and a good thing too. Since a diversion is assumed to be demanded after a shot at landing which has proved out of the question, diversion tables have overshoot, climb to a suitable level, reserve fuel all included in the figures; to peel off on the way down and head for the alternate field must require suitable corrections to these figures. In Data sheets 33 these corrections are clearly shown, but in sheets 34 they are not, and circumstances will decide which of the more thumbed tables are appropriate in the latter case, not forgetting the Low Level Cruise set.

Perhaps it's opportune to take a look at Data sheets 34; these are geared for heavier aircraft, but are similar in format, and mainly self-explanatory, though watch the footnotes as before. Interpolations for fuel consumption between stated weights are definitely only to the nearest 100 kg, and the Low Level cruise table is the one to use if holding.

An aircraft cruising at 0.83 indicated Mach at FL 350 is on Track to destination B which is 640 nm distant. Aircraft weight is 238 500 kg, temp dev +7°C, headwind component 40 kt.

Later, descent on Track is commenced, headwind component 20 kt.

(a) Give the time and fuel required for:
  (i) Cruise
  (ii) Descent

The aircraft arrives at B, but diverts after an overshoot to D, 156 nm distant, tailwind component 30 kt, temp dev +8°C.

(b) Give the aircraft weight at commencement of diversion.

(c) Determine time, flight level, and fuel required for diversion.

The aircraft holds over D at FL 160 for 17 min, temp dev +14 (Table 34D).

(d) Give the fuel required for holding.

Descent 15 min, 2 000 kg, TAS 378 kt, from Table 34E
∴ G/S 358 kt, distance run 90 nm.

Cruise which will be for 550 nm
TAS 484 kt, ∴ G/S 444 kt, time 75½ min

AUW 238 500 kg

Enter Table 34C, temp devn 0 to +9°C (page 10) at the FL 350 and guesstimate a fuel flow for the stage of 9 400 kg/h. The fact that it is just over an hour is not significant.

Cruise fuel = 75½ min at 9 400 kg/h = 11 830 kg

Answers (a) (i) 75½ min 11 830 kg
(ii) 15 min 2 000 kg

Now for the diversion; fuel used to B is 14 000 kg, and the aircraft weight after overshoot is 238 500 − 14 000 = 224 500 kg.

Entering Table 34B, interpolate for 156 nm and a 30 kt tailwind, read fuel 13 900 kg, FL 220, time 31 min. The start of diversion weight is 30 500 kg less than tabulated, so footnote correction (a) must be applied; this is 6% of 13 900, subtractive, 800 kg, = 13 100 kg.

Answer (b) 224 500 kg
(c) 31 min, FL 220, 13 100 kg.

For the hold, TAS 434 kt is extracted from the Table 34D, but the fuel flow must be checked against aircraft weight as per the footnote. Diversion started at 224 500 kg AUW, and 13 100 kg was to be used; this figure contains 7 500 kg reserve, and the aircraft has descended only to FL 160. Of the actual fuel required, 5 600 kg (13 100 – 7 500), the amount unused at the hold point would be the descent from FL 160 to landing, a figure of 1 800 kg from the descent table. A round estimate of what fuel has in fact been burnt off would be 3 800 kg, and the AUW at holding 220 700 kg. Thus, the noted consumption of 12 500 kg in the Table is satisfactory, and the correction element is not applicable.

Answer (d) 17 min at 12 500 kg/hr = 3 550 kg.

This example has put you into the diversion table, but if the descent had been broken off, say, at FL 180, whence to proceed direct to the alternate, then the calculations must be made from the descent, climb and cruise tables, starting from the AUW at the time of break off; since the diversion table showed that FL 220 would be climbed to, then from FL 180 a climb to around FL 340 would be possible and advisable.

Just to make sure you're not betting on avoiding a question on 3-engine operation, take a look at Table 34G in Data sheets 34 for such a problem as the following.

An aircraft en route to K goes on 3 engines at 1307.

Descent will be made on Track. Details are:

Distance	926 nm
Cruise wind comp	–33 kt
Descent wind comp	–36 kt
FL	310
Temp dev	+8°C

Aircraft weight at 1307 is 235 700 kg

Fuel in tanks at 1307 is 41 300 kg

(a) Give ETA K
(b) How much fuel remaining on landing?

Descent first: TAS 358 kt, fuel 1 900 kg, time 14 min
∴ G/S 322 kt, distance 75 nm, and cruise distance is then 851 nm.

AUW 235 700 kg, mean TAS 451 kt, 9 700 kg/hr at first, from the appropriate section of Table 34G. (Mean weight will be about 230 000 kg in first hour)

∴ G/S 418 kt, time 2 hr 02 min

Fuel		
	first hour	9 700 kg
	next 62 min	9 500 kg (use consumption 9 200 kg/h)
	Cruise fuel	19 200 kg
	+Descent fuel	1 900 kg
	Total used	21 100 kg

This total subtracted from fuel available at 1307 gives 20 200 kg left on landing.

ETA 1307 + 2 hr 02 cruise + 14 min descent = 1523 hrs

Answer (a) ETA K 1523
(b) 20 200 kg.

Quite straightforward, providing you have familiarity. As a rider, the 3-engine cruise in our favourite Data Sheets 33, Table 33H, is set out page by page for temp dev from standard, giving TAS at height and consumption per hour for a given weight at start: descent would call for normal descent Table 33E. Take a look at it right now or you'll be sorry.

To sum up for the Flight Plan itself, and such matters just discussed, you will need to do some of the published exam papers to get up some reasonable speed with accuracy: there is no need to be pedantic about fuel consumption. For instance, the tables themselves are not precise to a couple of hundred kilos – a weight of 101 100 kg gives a consumption 6 400 kg/hr and the following hour the AUW at start is 95 000 kg. There is a lot of averaging out, and though precision is always to be aimed for, it must be reasonable. You will find too the Drift and Wind Component table at the end of the book helps speed things along, using the appropriate table for the climb or level: all that computer work is avoided. The failing point in Flight Planning is pure arithmetical error, frequently induced by examination neurosis.

Flight Planning examinations are now in the multi-choice format. If you find it difficult to envisage how this can be done you should obtain CAP 505 or CAP 519 (see Appendix 3). You will find that it is still necessary to complete a Flight Plan as a preliminary to answering a series of objective questions.

9: Choice of Route

On an airline running scheduled services, it would appear at first sight that the Captain has precious little say: certainly the majority of local trips around the UK to the European continent are fixed on an airways route at pre-arranged altitudes, and fortunately for pilot morale, however much they may appear to resemble a taxi service, the vagaries of weather and the need to practise all types of let-downs are ever present. On long routes, despite the firm establishment of various different tracks across the water or desert, the Captain must study the overall weather picture at selected heights and pick the best route for speed, the best height for his particular aircraft under the conditions, never forgetting passenger comfort (or animal comfort if he's carrying a load of monkeys), viewing the whole thing with an eye on fuel consumption and safety at all times. This takes some expertise to do briskly and surely, and while there is nothing worse than the type who hums and ha's muttering 'ye canna be too careful', it is positively better than the impulsive one who decides too quickly and pours his 100 ton flying cigar into turbulent weather away from operating navigation aids.

The scheduled services are but a part of the Airline picture: any number of firms specialise in charter operations, and the majority are prepared to do charters, hiring aircraft if necessary from their competitors. Immediately, the profit motive could incite the Captain to take undue risks, especially if he is recently promoted to command and is anxious to make a name for himself as a good company man. Happily, by the time he is ready for such elevation, he has learnt more sense, apart from the legislative exercises he has had to undergo.

In such operations, the route and height are his decision: he will have in good time pondered the variables, and be ready at the Met briefing with a selection of possible routes from which to make a quick and safe choice; in fact, he may already have decided from his bedside after listening to the met man and the operations chap, so that on arrival at the field the Flight Plan is prepared and he needs only to check and corroborate that the latest information confirms his previous telephone briefing.

The procedure hardly varies; knowing his aircraft's heights for optimum operation, power- and fuel-wise, he will view a route first which will give him the best time track, examine it for trappy forecasts of turbulence or icing; for nav aids en route; for active danger areas notified for the time on the Notams; for Air Traffic Control restrictions and requirements; for safe clearance of topographical obstacles. Can he get over the highest mountains en route at the weight he will be at the time he gets there? Not only over them, but well over them? The broad decision now taken, he must at once examine the forecast weather at destination and departure field and at suitable alternates; not only alternates at his destination, but at the departure point, in case of return. Is there an en route aerodrome available for landing if the destination clamps, thereby avoiding a possible diversion to some destination alternate when fuel is getting low, and the destination alternate is suffering from the same foul weather as the destination itself? Is a chosen alternate not only far enough away from the clamped destination to be reached comfortably with the fuel aboard, forecast OK for weather, but also politically OK for the passengers and crew to be allowed through immigration in the case of a long wait? Is the required type of fuel available there? Are the takeoff and landing conditions restrictive? Are the necessary services available there at the possible arrival time? The world is scattered with airfields which do not fill all these requirements, only useful in case of *force majeure.*

The next check is on TOW and landing conditions: at expected TOW will the met conditions allow a safe unstick? With that TOW, less the expected fuel consumption from departure to destination (Item A + oil and water used, + extra required for climb, taxi, T/O, circuit and landing), is the maximum landing weight greater than the maximum allowed for the aircraft or by the airfield itself? If so, will the fuel uplift be reduced to allow a safe operation? Or should the payload be reduced?

He can now address himself to cruise control and fuel: long range or high speed cruise, depending on whether fuel conservation is more important than speed, or whether speed is possible with no fuel problems. All aircraft manufacturers produce their tables, and a little experience of their operation makes the decision more or less immediate. With the Item A + fuel required for alternate (latter usually at Long Range Cruise) he now considers his reserves, bearing in mind all the previous factors mentioned. A Route Contingency reserve is usually laid down by the Company, a percentage of Item A, with a maximum amount: this allows for the hard trip when actual winds are more adverse than forecast or for any of those happenings which are part of the flying game, such as being ordered to fly at an unsatisfactory altitude for the aircraft, or to move off Flight Planned Track for any reason, weather or traffic. The amount of contingency fuel is normally determined by the route: over country plentifully supplied with good airfields, the percentage of Item A would not be so large as that over the oceans or deserts. A similar percentage is usually applied to the alternate fuel, and for the same reasons. An emergency reserve is frequently added for Mother, + a goodly quantity for stand off, climb out, and taxi, the amount depending on the aircraft, and the complexity of the traffic at destination. It is almost normal in dodgy weather to have a stack of 20 aircraft at a place like NEW YORK, and plaintive cries from pilots that fuel is low and precedence is required are viewed very palely indeed from the other poor devils holding at precise altitudes for hours on end.

The only likely major difference to this type of routine will be if the destination is an island set solitary-like in the silver sea, a hearty distance from another aerodrome. Then, once having passed the Point of No Return, or the latest time to divert to a suitable field on the beam of Track, and a landing at destination becomes obligatory, an Island Reserve is substituted for alternate fuel, reserve fuel, and stand off fuel, to permit a long hold.

Add the lot up, and that's the Load Sheet Fuel: an Endurance is worked out from this from a graph or a rule of thumb average to give the maximum time the aircraft can be airborne.

Sundry wrinkles will become apparent, nearly all allowed for in the Aircraft Type Manual. The total fuel on board may include a quantity of unusable fuel in the tanks:

the only interest in this for the operation is that it's part of the weight. Climb, taxi, take-off fuel will be laid down in the Manual, and included on the Flight Plan; en route climb and allowances for it will be considered in the body of the Flight Plan from an appropriate table or graph; fuel for heaters, de-icers and so on are similarly allowed for. One pretty point often overlooked especially on shortish sectors is to jug up to the gills with fuel where the price is cheap, or to take the minimum consonant with safety where it is high: this will endear you to the commercial side of the company, for the savings can be appreciable. However, companies will usually lay down very precise rules as to when excess fuel is to be loaded.

10: Weight Calculation

Being in an international business the pilot is constantly plagued with units different from the ones he's been brought up on, and despite efforts to bring them all to one acceptable type world-wide, there's always the nation which won't conform or won't agree. In general, kilograms are becoming the accepted weight unit, though the pilot will find pounds aplenty on the trips. Volume should thus be in litres, and this comes hard to many, used to Imperial gallons. The U.S. gallon is only about 4/5 of the Imperial gallon, so there's another snag. It is quite unnecessary to memorise the conversion units, they're all on the computer anyway or in the Flight Manual, but when dealing with large figures, you should have an approximate idea of the relationship in order to get the number of noughts correct.

1 kilogram is 2·2 lb
1 litre is about 1/5th Imperial gallon
1 litre is about 1/3rd of a U.S. gallon

The weight of fuel varies with temperature and air pressure: the conversion from volume occupied (i.e. litres or gallons) to weight (kg or lb) is found by knowing the fuel's specific gravity at the time of loading. An engineer will have used his hydrometer to find this, and the sum is simple. It must of course be entered on the Load Sheet; on the Flight Plan, weight is the only concern.

The specific gravity is simply the relation of the weight of fuel at the time for a given volume to the weight of water for the same volume (the water being under standard conditions of temperature and pressure).

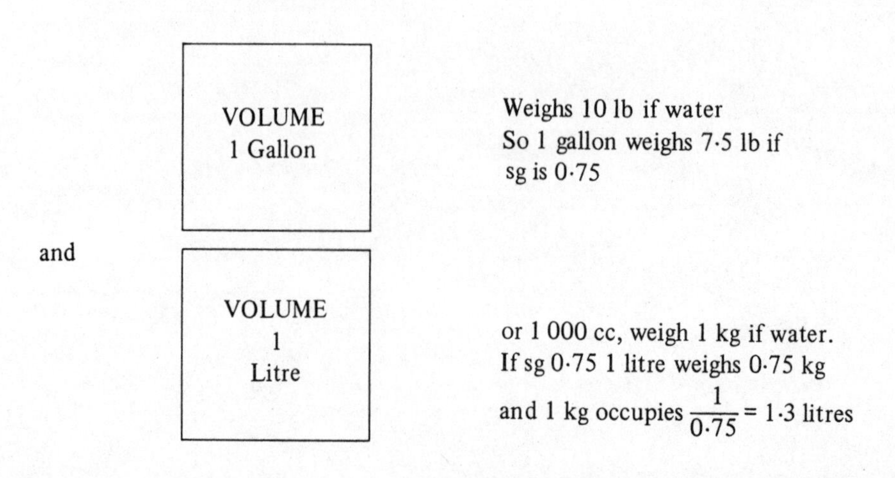

Fig. 10.1

The circular slide rule works all this out for you, and you will see that kg to lb is straightforward, but you cannot convert litres to kg, gals to lb, or any variant of these without knowing the sg – the errors can be considerable, and there must be no guesswork at all.

The precautions to be observed with respect to maximum TOW and maximum landing weight after obtaining total fuel requirements have already been mentioned; the fuel requirements, although calculated with precision, are the minimum requirements for safe operation, for there would be no point in lugging excess fuel around; thence, the payload carried must be such that these maxima are not exceeded, and off-loading passengers or freight is a serious decision in a commercial concern. But just as maxima are laid down for aircraft weight, so for each flight there must be a maximum payload that can be carried.

Consider the following example:

Maximum TOW 250 000 kg
Maximum Landing Weight 190 000 kg
Weight without Fuel or Payload 170 000 kg
Fuel on Board 23 535 kg
Fuel required from departure to destination 16 535 kg

The point to start with is Max Ldg Wt 190 000 kg. The only difference between this imperative maximum and the actual TOW is the fuel used up from departure to destination, the 'burn-off'.

So 190 000 kg
16 535

206 535 is the TOW

This is well below the maximum TOW, but dare not be exceeded, for if it was, the aircraft would be above maximum landing weight at the destination and would be forced to chunter around simply to use fuel and get the weight down.

The weight without fuel or payload, 170 000 kg, may now be added to the total fuel on board to give 193 535 kg, the weight without payload. Then 206 535 – 193 535 = 13 000 kg payload, pretty poor for such a heavy aircraft, but when going to spots like Iceland, calling for much fuel for an alternate in Scotland, such a case can frequently happen.

Another then:

Maximum TOW 47 300 kg
Weight less fuel and payload 33 400 kg
Fuel required from departure to destination 9 775 kg
Reserve fuel (assume unused) 1 985 kg

What is maximum payload that can be carried?
47 300 – 33 400 = 13 900 kg = fuel + payload.
FOB is 9 775 + 1 985 = 11 760 kg
∴ Payload is 2 140 kg

All the problems boil down to either of these types, and in practice the Station Duty Officer has a simple form to resolve them. The 'burn-off' may include not only the fuel used from departure to destination but also oil and water.

Regulated landing weight (RLW)	52 618 kg	Max TOW under forecast
Burn-off	18 240	conditions 72 575
	70 858 kg	
Max Permissible TOW	70 858 kg	
Estimated weight, no fuel (Empty Tank weight)	42 628	(including traffic load)
Max Fuel Uplift	28 230 kg	
Flight Plan Fuel	28 230 kg	
Excess Available	NIL	
Loadsheet Fuel	28 230 kg	
+ Taxi and etc.	500	
This is the fuel in your tanks	28 730 kg	TOW 70 858 kg

And you will see that the restricting factor on this trip was landing weight, and the fuel uplifted exactly the Flight Plan requirement.

Zero Fuel Weight (ZFW)

There is one further restriction – the Maximum Zero Fuel Weight (MZFW). Modern aircraft carry most, if not all, their fuel in the wings. If the tanks are empty, there will be a maximum permissible weight for the aircraft including all its contents (equipment, crew, passengers and cargo). Exceeding this weight would put an unacceptable load on the aircraft structure. To check that all three restrictions are complied with, the following procedure, illustrated by an example, is recommended.

Example 1

RTOW 167 t (tonne = 1000 kg) (Regulated Take-Off Weight)
RLW 139.55 t
MZFW 132 t

Aircraft weight without fuel and payload 90 t (often called the Aircraft Prepared for Service weight – APS weight)

Reserve fuel	3 t
Flight time	3 h 33 min
Diversion time	51 min
Fuel flow throughout	2.5 t/h

What is the maximum permissible TOW and payload?

Solution

TOW		LW		ZFW		
		139.55 t		132 t		
	Burn-off	8.88 t	Total FOB	14 t		
167 t		148.43 t		146 t		

Burn-off	8.875 t
Diversion	2.125 t
Reserve	3.000 t
Total	14.000 t

Max Permissible TOW is lowest of the three figures = 146 t

APS weight	90	
Total FOB	14	104 t
maximum payload		42 t

Example 2

APS weight	40 t
RLW	49.55 t
RTOW	65 t
MZFW	48 t
Burn-off	12.89 t
Reserve	1.65 t

Calculate the maximum permissible TOW and payload.

Solution

TOW		LW		ZFW		
		49.55		48	Burn-off	12.89
	Burn-off	12.89	Total FOB	14.54	Reserve	1.65
65		62.44		62.54	Total	14.54

Maximum permissible TOW		62.44
APS	40	
Total FOB	14.54	54.54
Maximum payload		7.9 t

Example 3

Load sheet reads:

A/c wt, no fuel, no payload	63 200 kg
Max TOW	99 000 kg
Route fuel (excluding reserve)	18 200 kg
Reserve (assume unused)	3 000 kg

If max Ldg wt is 76 500 kg, and MZFW 74 000 kg, find:

(a) TOW when maximum payload is carried.
(b) Maximum payload.

Solution

TOW		LW		ZFW		
		76.5 t		74 t	Burn-off	18.2 t
	Burn-off	18.2	Total FOB	21.2	Reserve	3
99		94.7		95.2	Total	21.2

Maximum permissible TOW		94.7	(a)
APS weight	63.2		
Total FOB	21.2	84.4	
Maximum payload		10.3 t	(b)

Example 4

You are to fly from P to Q where your fuel is not available, and return to P: a maximum payload is to be off-loaded at Q, and a maximum payload uplifted there. The following are the pertinent data:

Distance P to Q	610 nm
Wt, no fuel, no payload	36 500 kg
Max Ldg wt	52 400 kg
Max TOW	63 000 kg
Reserve (unused)	4 000 kg
Fuel for each flight (circuit, take off, etc.)	500 kg
Mean consumption	1 350 kg/hr
Mean G/S P to Q	240 kt
Mean G/S Q to P	280 kt

Give (a) The fuel which must be uplifted at P.
(b) Maximum payload which can be carried from P to Q.
(c) Maximum payload which can be carried from Q to P.

Answer

Both flights use less fuel than MTOW – MLW, so MLW is restricting.
P to Q: 610 nm @ 240 kt = 2 hr 32 min @ 1 350 kg/hr = 3 420 kg fuel.
Q to P: 610 nm @ 280 kt = 2 hr 10 min @ 1 350 kg/hr = 2 925 kg fuel.

∴ FOB at P	3 420 kg	(P to Q)
	2 925 kg	(Q to P)
Reserve	4 000 kg	
Circuit	1 000 kg	(500 for return to P)
Fuel reqd.	11 345 kg	. . . (a)

P to Q

Wt, no fuel, no payload	36 500 kg	Max Ldg wt	52 400 kg
FOB	+11 345 kg	Fuel used	3 420 kg
		Circuit	+ 500 kg
Wt, no payload	47 845 kg		
		TOW	56 320 kg
		Wt, no payload	–47 845 kg
		PAYLOAD P to Q	8 475 kg . . . (

Q to P

Wt, no fuel, no payload	36 500 kg	Max Ldg wt	52 400 kg
Q to P fuel	2 925	Fuel used	2 925 kg
Reserve	4 000	Circuit	+ 500 kg
Circuit	+ 500	TOW	55 825 kg
Wt, no payload	43 925 kg		

∴ TOW	55 825 kg	
Wt, no payload	–43 925 kg	
PAYLOAD Q to P	11 900 kg	. . . (c)

11: Point of No Return

PNR is the point beyond which an aircraft cannot go and still return to its departure field within its endurance.

This is entirely a fuel problem, and some reserve for holding or diversion should always be allowed for before setting about the calculation. A PNR is scarcely pertinent on trips over land well served with airfields, though a pilot will often prefer, if his destination and destination alternates are forecast en route to be below limits for his ETA, to return home rather than lob into an airfield where conditions for waiting with a crowd of passengers are miserable, expensive or politically troublesome. But over the oceans and deserts, a PNR is a must; the time to it is noted on the Flight Plan, and the ETA thereat put in on departure: it can be amended on the way if forecast winds are diabolically different from actual, or the flight is conducted at a different height or power than planned.

The solution of the problem can be found by formula, simply solved on the computer. The distance to the PNR is the distance to be covered back if the aircraft returns, i.e. distance Out = distance Home. The time for this distance at Ground Speed Out plus the time for this distance at Ground Speed Home will equal your endurance time excluding reserves.

If E = total endurance in hours (excluding reserve)
T = Time to PNR in hours
O = G/S Out.
H = G/S Home
R = Distance to the PNR.

Then:

$$E = \frac{R}{O} + \frac{R}{H}$$

$$EOH = R(O + H)$$

$$R = \frac{EOH}{O+H}$$

and since $T = \dfrac{R}{O}$

$$OT = \frac{EOH}{O+H}$$

$$T = \frac{EH}{O+H}$$

Work in minutes, if you like, as the computer work is eased; and beware of assuming that a wind component Out of +20 must give a Wind Component Home of

–20; at lower G/S, drift is greater, so check the G/S out and home against Track out and home. Having obtained the Time to PNR, the distance can be readily found at G/S out, e.g. endurance 4 hours, excluding 45 min reserve, Track 300(T), W/V 270/40, TAS 200 kt

∴ G/S Out 164 kt G/S Home 234 kt

$$T = \frac{240 \times 234}{164 + 234}$$

$$= 141 \text{ min}$$

and 2h 21m at 164 kt = 386 nm

All straightforward and the accuracy of the result can be checked – 2h 21m out + 386 nm at G/S 234 kt or 1h 39m = 4h endurance.

Point of No Return on two or more legs

Weather systems and Traffic Control systems seldom permit a long drag on a Single Track nowadays, and finding the PNR on a route where one or more changes of Track are involved is quite simple, and rational, even with the changing flight conditions such as a return with one engine failed.

Example

Following are route details: ignore climb and descent:–

	Track(T)	Distance	W/V
TAIPEH – KAGOSHIMA	042	606	260/110
KAGOSHIMA – SHIZUOKA	064	417	280/80
SHIZUOKA –TOKYO	011	61	290/50

ETD TAIPEH 1020 GMT
TAS 410 kt (4 engines) 350 kt (3 engines)
Fuel Consumption 3 000 kg/hr (4 engines) 2 800 kg/hr (3 engines)
Reserve (assume unused) 45 min
Fuel on Board 15 000 kg

Give distance to and ETA at the PNR assuming the return flight will be on 3 engines.

Solution

The method used is a very powerful one that can take account of any number of variables. Although it may seem a little laborious, it should be remembered that nearly half the work will have already been done when the Flight Plan was prepared. Briefly the technique is:

(i) Progressively check the total fuel required to go out to and back from each successive turning point.
(ii) Eventually the fuel available will not be sufficient to go out to and return from a turning point. From the previous turning point decide how much fuel is available to go from there to the PNR and back to the point.
(iii) Comparing this amount with the total fuel required to go out and back on the

complete stage will give the ratio of distance and time out to the PNR compared with the stage distance and time out.

The importance of a systematic layout cannot be over-emphasised:

Outward flight: TAS 410 kt Fuel consumption 3 000 kg/hr

	Tr °T	G/S kt	Dist nm	Time min	Fuel kg
TAI–KAG	042	491	606	74	3 700
KAG–SHI	064	472	417	53	2 650
SHI–TOK	011	399	61	9½	480
Return flight: TAS 350 kt Fuel consumption 2 800 kg/hr					
TOK–SHI	191	354	61	10½	490
SHI–KAG	244	282	417	89	4 150
KAG–TAI	222	257	606	141½	6 600

Fuel analysis:

	Out	Back	Out + Back	Total
TAI–KAG	3 700	6 600	10 300	10 300
KAG–SHI	2 650	4 150	6 800	17 100
SHI–TOK	480	490	970	18 070
	6 830	11 240		

6 830
11 240
18 070 kg — Check that this figure agrees with the running total (underlined) above.

Fuel on board	15 000 kg
Reserve 45 min @ 2 800 kg/hr	2 100
PNR fuel	12 900
Fuel TAI–KAG–TAI	10 300
Fuel KAG–PNR–KAG	2 600
Fuel KAG–SHI–KAG	6 800

$$\frac{2\,600}{6\,800} = \frac{d}{417} = \frac{t}{53\frac{1}{2}}$$

d = 159 nm t = 20

	Dist	Time
TAI–KAG	606	74
KAG–PNR	159	20
TAI–PNR	765	95

ETA at PNR = 1020 + 1 h 34 min = 1154

Graphical Solution

Given: FOB 750 gal, TAS 180 kt, Consumption 95 gal/hr
Headwind component 25 kt
Find the PNR, leaving 50 gal in reserve.

Steps: (i) Endurance for 700 gal at 95 gal/hr = 442 min
(ii) Distance OUT for 442 min = 442 min at G/S 155 kt
= 1 142 nm
(iii) Distance HOME for 442 min = 442 min at G/S 205 kt
= 1 505 nm
(iv) With coordinates fuel and distance, plot these curves (Fig. 11.1).

The point of intersection is the PNR

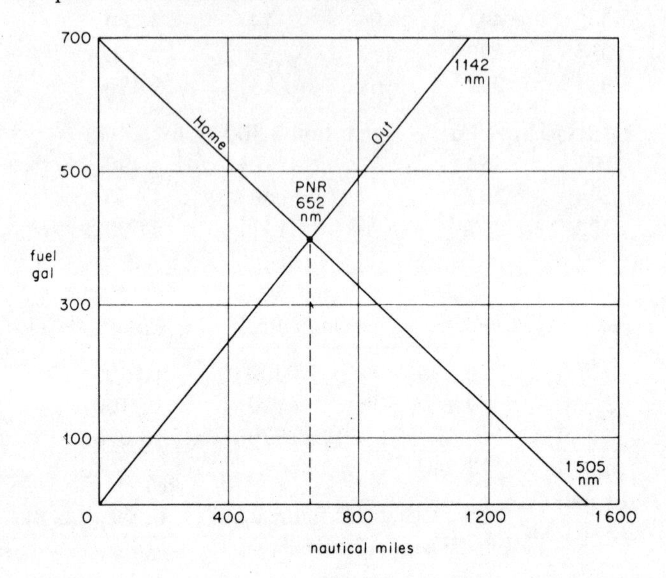

Fig. 11.1

This can be checked correct with the formulae. On the graph, as large a scale as possible should be chosen to ensure an accurate result.

Progress Charts (Howgozits)

To find a PNR graphically as part of the usual flight involving various FLs, W/Vs, speeds, etc., will necessitate the use of a progress chart or, as the Americans say, a Howgozit. The basic idea behind this is to present the pilot with a graphical forecast of how the fuel will be used during the flight. The graphs may be drawn upwards as in Fig. 11.2, or downwards (Fig. 11.3) starting with the total fuel on board and showing the fuel expected to be left at each turning point. This chart can then be used to obtain the PNR and also the Critical Point as explained in the next chapter.

Proceed as follows:

(i) Complete the Flight Plan.
(ii) Plot the Fuel/distance Out, starting with TOC and ending with TOD, reading the stages off the Flight Plan. This is in fact the chart to be used for the Flight progress; as the turning or reporting points are reached, the fuel used so far is entered on the graph, and a comparison with fuel planned is at once visually apparent.

(iii) Draw a line across the graph to represent the fuel available for PNR calculation; where this line meets the fuel coordinate may now be deemed the 'departure point' whither the aircraft is returning having used the PNR fuel.

(iv) From this point, work backwards to TOD and thenceforth in fuel used per sector, until the curves cross. Fig. 11.2.

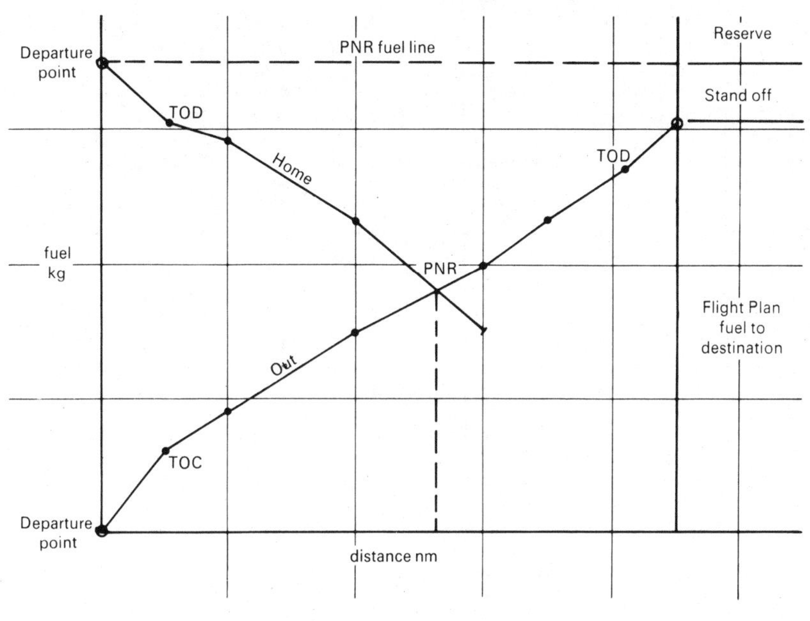

Fig. 11.2

Practical Solution of PNR on Progress Chart

In practice, it is usual to approximate the return flight, when obtaining the PNR, assuming one speed and one consumption throughout. Here is an example for a flight from Z to T via B, R. A and E worked out for you (Fig. 11.3)

Example

Assume:

(i) Outward flight of 2190 nm taking 260 min as plotted on the chart – derived from the Flight Plan.

(ii) Average return TAS 496 kt.

(iii) Average return wind component −13 kt.

(iv) Average return fuel flow 10 500 kg/hr. PNR reserve is 15 000 kg.

Solution

Choose a convenient return time about $\frac{3}{4}$ of flight time – in this case choose 3 hr.

3 hr at 10 500 kg/hr	=	31 500 kg at G/S 483 kt = 1 449 nm
PNR reserve		15 000
Total return fuel		46 500

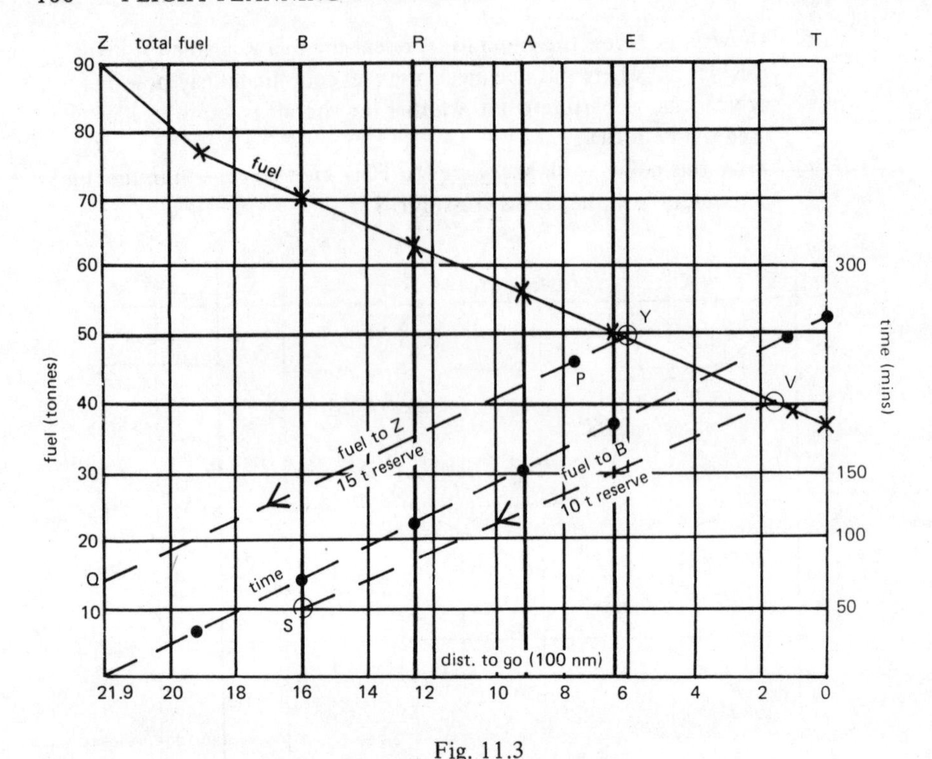

Fig. 11.3

Plot 46 500 kg at 2 190 – 1 449 = 741 nm to go (P).
Plot 15 000 kg at 2 190 nm to go (Q).
Join and extend QP to cut outward fuel graph at Y.
Y is the PNR – 600 nm to go to T.
Dropping a perpendicular down to the time line will give the time to reach the PNR from Z as 193 min.

The reasoning behind the method is as follows. At any point, for example A, the fuel to return to Z is represented by the distance from the outward fuel graph upwards (at A about 31.5 t). Similarly the distance downwards represents the fuel to return to Z with a 15 t reserve (at A about 42 t). The distance between the two graphs at any point (at A about 15 t) represents the fuel surplus to the requirement to be able to go out and still be able to return with 15 t available to Z. At Y this surplus has disappeared and so the requirement can only just be satisfied. Y is, therefore, the PNR required.

The great advantage of this technique for finding the PNR is its extreme flexibility. For example, if it was required to find the PNR for a return to B retaining a reserve of 10 t, it would be done by drawing a line through the 10 t point at B (point S) parallel to QP so that it cuts the outward fuel line at V at 170 nm to go. Time to V, 204 min.

Practical Significance of the PNR

Knowledge of the PNR is useful in cases where there is some doubt about the availability of the destination and its alternates. The most likely reason for this is weather but there are a number of other reasons why airports may be closed to traffic. One that could well affect all airports in a country is political unrest. Obviously before reaching the PNR, the Commander must make his decision as to whether or not to proceed with the flight. After the PNR, he is committed to carrying on to the destination or an alternate to it.

Engine Failure PNR

The common practice of assuming that the return flight from the PNR will be with one engine failed is designed to produce a PNR that is valid even if, after deciding to return at the PNR because of some emergency at the destination, the engine then fails with a resulting loss in economy on the return flight. It must be emphasised that the primary emergency is still that at the destination. If the failure of the engine was the primary emergency, it is more likely that the main point of concern would be: How soon can a safe landing be made? This is dealt with in the next chapter on the Critical Point.

Significance of PNR Reserves

Any PNR calculated must, inevitably, be inaccurate. There are so many possible causes of error – W/Vs, temperature, FLs, route, aircraft performance – to mention just some of them! The safeguard is in the reserves allowed. If these are sensibly adequate, any PNR should be valid, even if the assumptions made in the calculation are not accurate. The leeway that an adequate reserve provides enables many operators to rely on PNR tables for a particular aircraft. These are entered with take-off weight, fuel load and estimated average wind component for the operation and a distance to the PNR can then be extracted. Errors caused by the approximations involved should be taken care of by the reserves employed.

Factors Affecting the PNR

The maximum distance to the PNR will be achieved in still air conditions. In fact, a very quick approximation is to use half the still air range as the distance to the PNR. Any wind will reduce the distance for either or both of the following reasons:

(i) If there is any drift, the effective head wind component on a track is more than the effective tail wind component on the reciprocal track.
(ii) The aircraft will take longer and so experience more head wind effect (time x wind component) when flying into a head wind than it will experience tail wind effect on the reciprocal track.

The other two factors are simpler to understand:

(i) Fuel available. Other things being equal, the distance to the PNR will vary directly with the fuel available: 10% more fuel available will mean 10% greater distance to the PNR if nothing else, such as wind component, changes.
(ii) Aircraft performance. The better (more economical) the performance, the greater the distance.

Practical examination questions

Q1 If the PNR is calculated to be 880 nm with 10 000 kg of fuel available, the

distance to the PNR with 11 000 kg available, other things being equal, will be approximately:
(a) 928 nm (b) 968 nm (c) 960 nm (d) 920 nm

Q2 On a flight an aircraft is found to be achieving G/Ss 10% higher than planned. Assuming all conditions remain the same, the revised distance to the PNR will be:
(a) Less (b) More (c) Unchanged
(d) It is impossible to say

Q3 The distance to a PNR flying directly into a 50 kt head wind is 1 200 nm. The distance to the PNR with an exactly reciprocal W/V will be:
(a) Less than 1 200 nm
(b) More than 1 200 nm
(c) 1 200 nm
(d) It could be any value

Q4 With a TAS of 400 kt, the distance to a PNR in still air is 1 200 nm. The distance (nm) with a forecast W/V at 90° to the track of 40 kt will be:
(a) 1 194 (b) 1 206 (c) 1 140 (d) 1 200

Q5 The primary emergency for which a PNR with engine failure is computed for is:
(a) Engine failure
(b) Emergency at the destination and its alternates
(c) Any on-board emergency requiring a landing to be made as soon as possible
(d) Fuel shortage

Q6 In-flight checks reveal that the fuel flows are 4% greater than expected at pre-flight planning. If everything else is as expected, the distance to the PNR will be:
(a) 4% more (b) Unchanged (c) 2% less (d) 4% less

12: Critical Point

Critical point is the point from which it would take equal time to continue to destination as to return to a suitable aerodrome.

This is not a function of fuel: there is a critical point when crossing the road or swimming a river: distance and related G/S are the factors to consider and it is important to bear in mind that it is a Flight Plan problem initially, to prepare for some eventuality like an engine failure when an instant decision must be taken to proceed or return, the quicker being the choice as there is some concern among those present.

Again, the solution is done by simple formula, and the ETA CP entered on the Flight Plan; the same arguments hold as previously as to the trips on which a CP is vital.

Take a straightforward case first

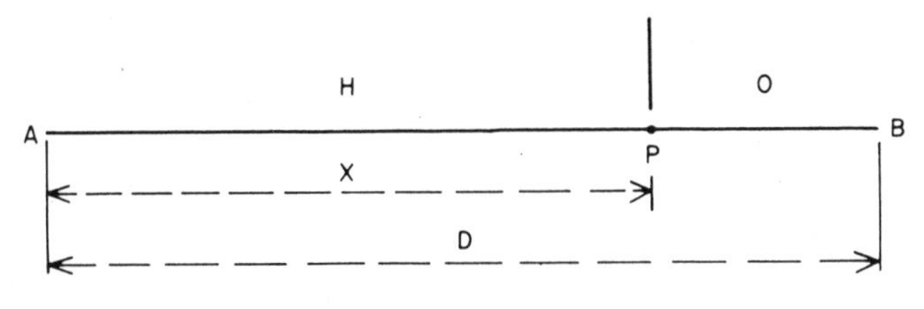

Fig. 12.1

Where D is total distance

P the Critical Point

X Distance to CP in nautical miles

O Ground Speed Out in knots

H Ground Speed Home in knots

Then by definition:

P to A at G/S Home = P to B at G/S Out.

i.e. $$\frac{X}{H} = \frac{D - X}{O}$$

$$OX = H(D - X)$$

$$OX = HD - HX$$

$$OX + HX = HD$$

$$\therefore \quad \frac{DH}{O + H} = X, \text{ the distance to the CP.}$$

Now the CP is bursting with importance when the aircraft is acting up, usually an engine out, not in itself an emergency, but leading towards it if something else happens: an aircraft on 3 engines will not go as fast as on 4, strangely enough, especially when fuel conservation is high priority. An operator, therefore, lays down in the manual an average 3-engined and 2-engined TAS at specified heights; thus, the CP data must be worked using the reduced TAS so that the equal times home and away from the CP are appropriate to the conditions should the exigency occur. In the air, once the CP is passed (and the ETA to it will be calculated at normal G/S, just like a reporting point), the pilot will proceed to his destination. A separate CP at full TAS can be calculated readily, to cope with serious situations like a loose panther in the hold, or a berserk and frothing passenger which affect the safety of the aircraft and its occupants, but not its power. But in a pressurisation failure, for instance, while the power is unaffected, the CP is dependent on a TAS at a new enforced height with implications very similar to the engine failure cases. This, too, calls for a separate CP, not an arduous calculation since the action for the pressurisation failure will be laid down and the optimum height with the appropriate data is set out in the aircraft manual.

There are several pertinent possibilities, then; and bear in mind that they are just that. One or more CPs are noted on the Flight Plan to be referred to as though they are turning points, with their ETA. Once a CP is passed, the pilot's action is clear: if a near-emergency arises, he will aim for the destination airfield. The CP is but a preparation in case of emergency, and if that emergency happens, he has the facts before him at once.

Some samples:

1. Track 240(T), W/V 310/35, TAS 260 kt Distance 530 nm
   ∴ G/S out = 245 kt G/S home = 270 kt

$$\text{Distance to CP} = \frac{\text{DH}}{\text{O + H}}$$

$$= \frac{530 \times 270}{245 + 270}$$

$$= 278 \text{ nm}$$

and time to CP = 278 at G/S out 245 kt = 1h 08m

Check

278 at G/S home 270 kt = 1h 02m
(530 – 278) at G/S out 245 kt = 1h 02m

2. What is distance to CP en route from DARWIN and MELBOURNE distance 1 728 nm, cruise TAS 425 kt, 3-engine TAS 400 kt, headwind component from CP to MELBOURNE 5 kt, headwind component CP to DARWIN 20 kt?
   ∴ for the CP calculation: G/S out = 395 kt, G/S home = 380 kt, } both at reduced TAS

$$\text{Distance to CP} = \frac{\text{DH}}{\text{O + H}}$$

$$= \frac{1\,728 \times 380}{395 + 380}$$

$$= 847 \text{ nm}$$

You check.

The ETA CP can then be found simply from the normal Flight Plan after departure; this type of problem is most frequently used in practice and, despite finding the wind components by inspection, is proved reasonably accurate: with a long trip going fairly to plan until an engine drops out, a pilot who turns back because it happens 5 minutes before the CP cannot be criticised for being dogmatically correct, but his employers and passengers might think him rather lacking in dash and élan.

3. Now for the several Track CP.

TAS 200 kt Engine failure TAS 160 kt

Route

BAGHDAD–BASRA	Track 115(T), Dist 170 nm W/V 180/20
BASRA–KUWAIT	Track 178(T), Dist 110 nm W/V 230/30
KUWAIT–BAHRAIN	Track 129(T), Dist 147 nm W/V 250/15

Find ETA CP if ATD BAGHDAD is 1115.

Using the reduced TAS to obtain the G/Ss, calculate the onward and return times.

	Tr	G/S	distance	time	
BGW–BAS	115	151	170	$67\frac{1}{2}$	
BAS–KWI	178	140	110	47	ONWARD
KWI–BAH	129	167	147	53	
BAS–BGW	295	167	170	61	
KWI–BAS	358	177	110	$37\frac{1}{2}$	RETURN
BAH–KWI	309	152	147	58	

Now prepare a diagram, as in Fig. 12.2, to show from each turning point the total time on to BAHRAIN and back to BAGHDAD. At the CP, the difference between the two totals will need to be zero. From inspection, it can be seen that a zero difference will be found between BASRA and KUWAIT. Interpolate between the time differences of −39 and +$45\frac{1}{2}$ to find the zero position:

$$\frac{39}{39 + 45\frac{1}{2}} = \frac{d}{110} = \frac{t}{37} \text{ (time at full G/S to go 110 nm)}$$

$$d = \underline{50} \quad t = \underline{17}$$

To find the total distance and the time from BAGHDAD (at full TAS):

	distance	time
BGW–BAS	170	$53\frac{1}{2}$ (170 nm /GS 191 kt)
BAS–CP	50	17
BGW–CP	220	$70\frac{1}{2}$

ETA CP = 1115 + 1 hr $10\frac{1}{2}$ min = $1225\frac{1}{2}$

Graphical Solution

A simple graphical solution of the following problem is shown in Fig. 12.3.

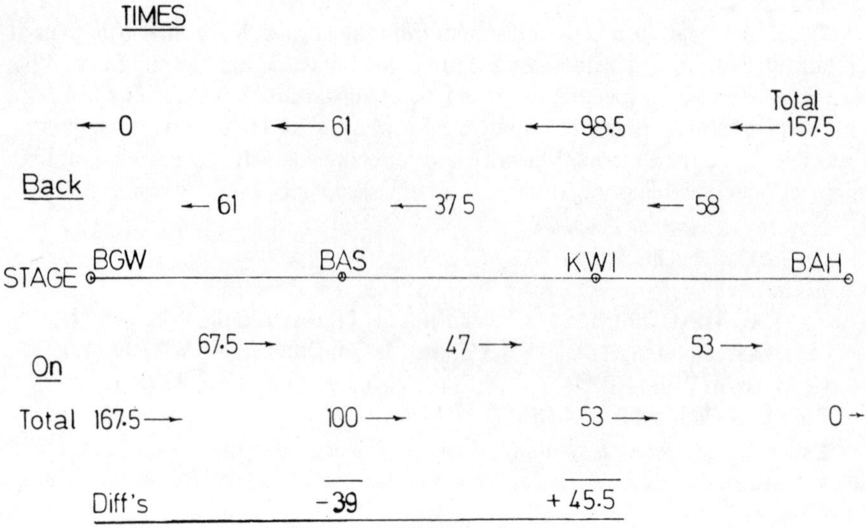

TIMES

	BGW		BAS		KWI		BAH
Back Total	← 0		← 61		← 98.5		← 157.5
Back		← 61		← 37.5		← 58	
STAGE	BGW		BAS		KWI		BAH
On		67.5 →		47 →		53 →	
On Total	167.5 →		100 →		53 →		0 →
Diff's			−39		+ 45.5		

Fig. 12.2

Example
Flight from A to B:
distance 850 nm; wind components: out −45 kt, back +40 kt;
full TAS 280 kt; engine failure TAS 240 kt.
Find the CP with and without engine failure.

Solution
Full TAS

	G/S	distance	time
A–B	235	850	217
B–A	320	850	159½

At X on the vertical line through A on the graph, plot the time to go to reach B and join to the zero time point at B. Similarly, plot Y on the vertical line through B the time to go to reach A and join to the zero time point at A. The intersection of the two lines at Z gives the CP at a distance of 490 nm from A.

Check by calculation:

Reduced TAS

	G/S	distance	time
A–B	195	850	261½
B–A	280	850	182

Follow a similar plotting procedure as indicated by the pecked lines in Fig. 12.3 and obtain a position for the reduced TAS CP of 500 nm from A. Again check by calculation. Notice that the CP has moved upwind with the reduced TAS. It is not possible to read off the time to the CP directly from these graphs although the

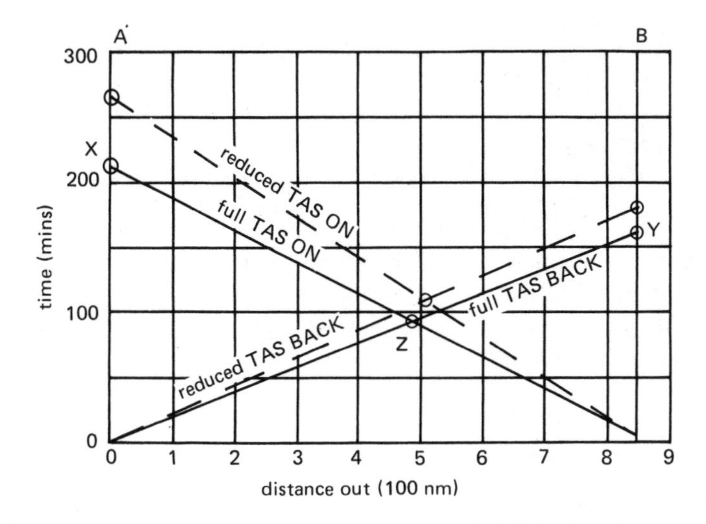

Fig. 12.3

necessary information could be obtained from the full TAS 'ON' graph by a process of subtraction:

Time on at the full TAS CP	=	92
Total time A to B at full TAS	=	217
Time to reach full TAS CP	=	125

To find the ETA for the reduced TAS CP is a little more complex and involves checking the distance to go on the full TAS 'ON' graph first. See if you can figure it out – check your answer by calculation.

Critical Point by using Progress Charts

In the previous chapter, it was seen that the PNR could be derived from the Progress Chart. The CP can also be obtained although, strictly speaking, it has no connection with fuel. The method is to find the point where the fuel needed to fly to the two aerodromes being considered is the same. As we are considering one particular aircraft, at the same weight and operating in a fairly consistent ambience of temperature and pressure, it would indicate that equal fuel represents equal time.

Figure 12.4 is a reproduction of Fig. 11.3 with the outward time graph omitted for the sake of clarity. To obtain the CP between Z and T, draw a line US from the zero fuel point at Z (point U) parallel to the return fuel line QPY. SU represents a return flight to Z to dry tanks (0 reserve). Similarly, draw in the line NW parallel to the outward fuel line ZYL. As ZYL is not a straight line, paralleling it is best done by using a pair of dividers, opened to the distance LW (the outward flight fuel reserve), to drop each plotting point on ZYL by the same amount. The intersection of NW and SU at 'C' will give the required full TAS CP. A reduced TAS CP would necessitate adjusting the slopes of the two dry tank fuel lines (SU and NW) for the change in economy at the lower speed.

The triangle UCW forms what is sometimes referred to as the danger zone. If in flight, plots of fuel remaining against distance to go show a trend of running into the triangle UCW, it would indicate a very serious situation. Within the danger zone neither Z nor T could be reached even by flying down to dry tanks!

Comparison of PNR and CP

These are quite often confused. The following distinctions should be understood:

(i) PNR is required when there is no possibility of a safe landing being made at the destination or its alternates. The CP is required against the possibility of an emergency in the aircraft requiring a landing to be made as soon as possible

(ii) Although the symbols O and H are used in both the CP and the PNR formulae, they have different meanings:

Symbol	PNR	CP
O	Full GS OUT from A to PNR	Full or reduced GS FROM CP ON to B
H	Full or reduced GS BACK from PNR to A	Full or reduced GS FROM CP BACK to A

Careful inspection will show that in a case with varying wind components throughout the flight and/or a requirement for engine failure to be considered, there could be considerable differences between the values used for the CP and PNR calculations.

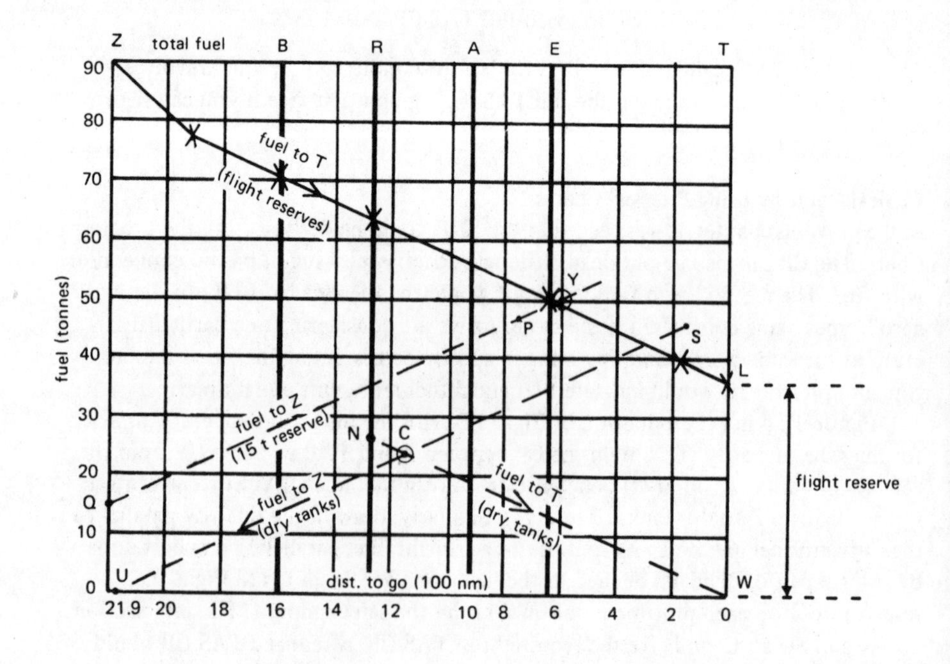

Fig. 12.4

(iii) PNR depends on endurance. It is quite possible, given sufficient fuel for the PNR, for the PNR to be at the destination. This just indicates that the reserves are sufficient to fly out to the destination and return without refuelling. To some isolated places or aerodromes where the price of fuel is very high, this may be a very sensible arrangement. On the other hand the CP is dependent not on fuel but on the total distance between the two bases. It will always tend to be around the mid-point area.

Relationship of CP and PNR

Having insisted on the distinctions between the CP and the PNR, let us now consider their affinity to each other. In a straightforward case with only one track, W/V and TAS to consider, the 'H' and 'O' values would be identical in the two formulae and so if the CP and PNR are to coincide:

$$\text{CP} = \frac{\text{DH}}{\text{O} + \text{H}} = \frac{\text{EOH}}{\text{O} + \text{H}} = \text{PNR}$$

so $\text{DH} = \text{EOH}$

or $\text{D} = \text{EO}$

or $\frac{\text{D}}{\text{O}} = \text{E}$

Distance divided by the G/S out is the flight time and so when the PNR endurance equals the flight time, the PNR and CP will coincide in this simple case. However, even in a more realistic complicated case of varying wind components and performance, the relationship can still hold good. In Fig. 12.5, a flight is due to reach the CP at 1500 and the destination B at 1800. By definition, if the aircraft turned round and returned to A at the CP, the ETA back at A must be identical with the ETA at B, i.e. 1800 in this case. So, if the aircraft in this case had a PNR endurance equal to the flight time (7 hours), the CP and the PNR would be coincident. Notice that no provisos have been put in about constant wind components or performance. It is, therefore, true in all cases not involving engine failure, that the CP and PNR will be coincident if the PNR endurance equals the flight time. This implies that the PNR reserves will be the same as those being carried for the flight outward. PNR reserves will normally consist of a holding reserve and a diversion reserve, whereas the flight outward reserves will probably include a contingency reserve also. As a result the PNR will usually lie beyond the CP although, if the return diversion is much greater than the destination diversion, this may not be true,

Factors Affecting the CP

In still air conditions or with a wind at 90° to track giving an equal effective head wind component in both directions, the CP will be exactly half-way between the two bases being considered. The factors affecting the position of the CP are:

(i) Total distance between the two bases.
(ii) The wind components – the CP will always be up-wind of the mid-point.
(iii) The TAS – reducing the speed will increase the significance of the wind and so cause the CP to move further along track into the wind.

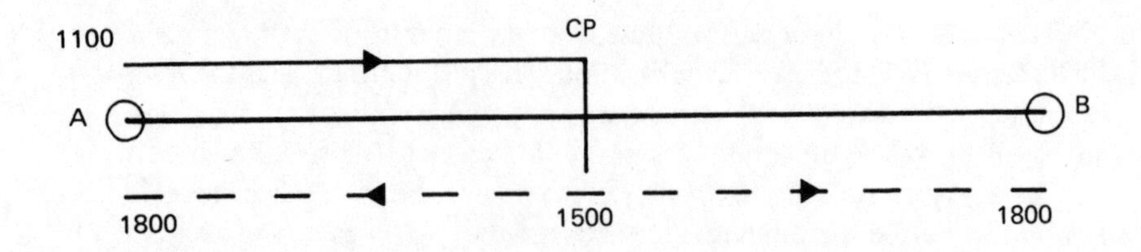

Fig. 12.5

Given a head wind component (W) going out and an equal tail wind component coming back, the CP formula may be written:

$$\frac{D(TAS + W)}{2TAS} = D\left(\frac{1}{2} + \frac{W}{2TAS}\right)$$

$$= \frac{D}{2} + \frac{DW}{2TAS}$$

This means that the CP will be displaced from the mid-point by a distance equal to:

$$\frac{DW}{2TAS}$$

Based on this some Operation Manuals provide a simple table for adjusting the half-way point to derive a Critical Point. Here is a typical example:

This table gives the percentage increase of the half-way distance necessary to allow for a 10 kt head wind component.

TAS	100	200	300	400	500	600
%	10	5	3.3	2.5	2.0	1.7

(note % = 1 000 ÷ TAS)

Taking the first example in this chapter, the approximate wind component was 12½ kt (15 kt out and 10 kt back) and the TAS 260 kt. The above table suggests about 4% adjustment for a 10 kt wind at a TAS of 260 kt. This represents 5% for 12½ kt. The total distance was 530 nm and so the half-distance is 265 nm; 5% of this is 13 nm. Adding, because it is a head wind (CP moves into wind) gives the correct distance to the CP of 278 nm.

Practical Multi-choice Questions

Q1 The effect on the position of the CP of reducing the TAS if there is a head wind component will be to:
(a) Increase the distance
(b) Decrease the distance
(c) Leave the distance unchanged
(d) Move it nearer the half-way point

Q2 In flight and flying at the planned TAS, it is found that the aircraft is achieving faster stage times than planned. Presuming the situation remains unchanged, the CP's position will be:

(a) Unchanged
(b) Nearer to the half-way point if tail winds were forecast
(c) Nearer to the destination
(d) Nearer to the departure point

Q3 A CP is calculated for a 1500 nm flight assuming a 50 kt head wind component out and a 50 kt tail wind component coming back and comes to 825 nm. It is discovered that the winds are the wrong way round. The correct distance to the CP will be:
(a) 675 nm (b) 825 nm (c) 750 nm (d) Some other value

Q4 In the event of a return to the departure point, 10 000 kg of fuel should be available when back over the departure aerodrome. The flight reserves being carried are 15 000 kg, the average fuel flow is 5 000 kg/hr, TAS 400 kt and there is a 'dead' head wind out of 100 kt. The position of the PNR in relation to the CP will be:
(a) 200 nm further
(b) 250 nm further
(c) 188 nm less
(d) 188 nm further

Q5 Compared with still air conditions, the CP with a strong wind at 90° to track will be:
(a) In the same position, with an earlier ETA
(b) In the same position, with a later ETA
(c) At greater distance, with the same ETA
(d) At shorter distance, with the same ETA

Q6 Flying across the North Atlantic with the usual Westerly winds the CP will be:
(a) Only nearer to N America when flying Eastwards
(b) Only nearer to N America when flying Westwards
(c) Always nearer to N America
(d) Always nearer to Europe

Q7 The effect on the CP of reducing the TAS is to:
(a) Always increase the distance to the CP
(b) Always reduce the distance to the CP
(c) Always move it along the track further away from the mid-point
(d) Have no effect in the case of zero winds or winds at 90° to track

Q8 To calculate distance and time to the CP in the case of engine failure use the:
(a) Reduced TAS for all calcuations
(b) Reduced TAS for all the distance calculation only
(c) Reduced TAS for G/S back but full TAS for G/S on
(d) Full TAS for all calculations

13: Flight Planning Re-Check

We have laid down the principles and methods of flight planning, but for all the pages a student will be faced with in the professional examinations, speed and accuracy are positively vital in this for satisfying the examiner. So plenty of practice is required in preparation – and there is not much time for cogitation when practically called on while flogging the routes either. Incidentally, polish up the old examination technique; allot the appropriate time to a question that its marks warrant. Divide the number of questions into the time allowed so that you can easily check if you are keeping up with the clock. Do not waste time on any unfamiliar or awkward question. Leave it and, hopefully, you will have time to return to it when you have dealt with all the easier questions.

CP/PNR invariably feature on the planning papers. When dealing with them it is important to remember that:

PNR distance out is the same as the distance home. Its position depends on the fuel available for its calculation.

CP time home is the same as the time to destination.

Example 1

An aircraft is to fly from A to B on a Track of 280T distance 959 nm, mean TAS 230 kt. W/V for the first 430 nm is 200/50, and 260/65 for the remaining distance. Fuel on board is 26 500 kg, 3 100 kg to be held in reserve: consumption 3 400 kg/hr. Give the time and distance to:

(a) PNR,

(b) CP, assuming engine failure at the CP and a reduced TAS of 190 kt.

Answer

(a) PNR. Draw a diagram first, and insert what is known: a rough direction is adequate, of course.

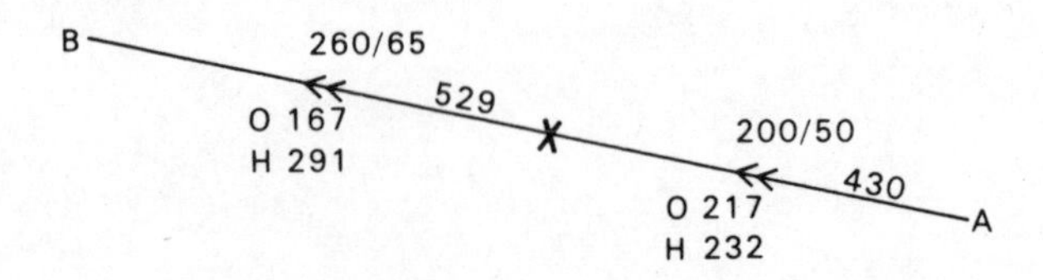

Fig. 13.1

Tr 280T
TAS 230 kt

Endurance = (26 500 – 3 100) kg @ 3 400 kg/hr
= 23 400 kg @ 3 400 kg/hr
= 413 min

Insert on the diagram the G/S for each leg evolved from the computer.
Treat as two legs, and try to lose A to X in the first place.

A to X out 430 nm @ 217 kt = 119 min
X to A home 430 nm @ 232 kt = 111 min

230 min

Thus, as the total endurance is 413 min, the PNR is beyond X, and we have endurance from X of 413 – 230 = 183 min for the calculation.

Formula

$$T = \frac{EH}{O + H} \quad \text{from X}$$

$$T = \frac{183 \times 291}{291 + 167} \text{ min}$$

$$= \frac{183 \times 291}{458} \text{ min}$$

$$= 116 \text{ min or } 1 \text{ hr } 56 \text{ min}$$

and at 167 kt = 324 nm from X.
Thus distance to PNR from A is (430 + 324) nm = 754 nm
Time to PNR from A is

	1 hr 59 min	from A to X
+324 nm @ 167 kt	1 hr 56 min	from X to PNR
	3 hr 55 min	from A to PNR

Should the time out from A to X, plus the time home from X to A come to more than the endurance of 6 hr 53 min, then the PNR is on AX, and the second leg is superfluous: the formula could be entered at once. You should be so lucky.

A quick check for correctness:

A to X	1 hr 59 min
X to PNR	1 hr 56 min
PNR–X (324 nm @ 291 kt)	1 hr 07 min
X to A	1 hr 51 min
	6 hr 53 min which is our endurance.

(b) CP (reduced TAS) (See Fig. 13.2)

		ON		BACK		ON (FULL TAS)	
	distance	G/S	time	G/S	time	G/S	time
A–X	430	176	147	192	134	217	119
X–B	529	128	248	250	127	167	190

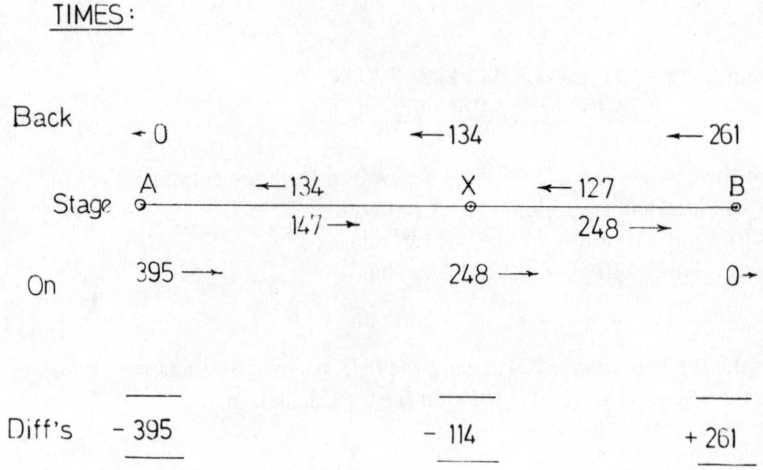

Fig. 13.2

From the diagram, CP lies between X and B and so:

$$\frac{114}{114 + 261} = \frac{d}{529} = \frac{t}{190}$$

$$d = 161 \qquad t = 58$$

	d	t
A–X	430	119
X–CP	161	58
A–CP	591	177 = 2 hr 57 min

Check

CP to Ḃ = (529 – 161) nm @ 128 kt = 368 nm @ 128 kt = 2 hr 52 min

CP to X = 161 nm @ 250 kt = 38 min
X to A = 430 nm @ 192 kt = 2 hr 14 min

2 hr 52 min

Example 2

This one is severely practical:

Given max TOW weight	61 000 kg
Weight, no fuel, no payload	37 000 kg
TAS	410 kt
Distance	2 250 nm
Consumption	2 800 kg/hr
Reserve (assume unused)	3 200 kg
Wind component outwards	–40 kt
Wind component back	+40 kt

Determine.

(a) Maximum payload that can be carried.
(b) Time and distance to the CP.
(c) Time and distance to the PNR.

Answer

G/S 370 kt, distance 2 250 nm.
∴ Time to destination 6 hr 05 min
At 2 800 kg/hr, fuel used = 17 035 kg, rounded off.

Empty wt	37 000 kg	Max TOW	61 000 kg
Fuel	17 035 kg	Wt, no payload	57 235 kg
Reserve	3 200 kg	PAYLOAD	3 765 kg . . . (a)
Wt, no payload	57 235 kg		

CP.

G/S out, 370 kt; G/S home, 450 kt.

$$\text{Distance to CP} = \frac{DH}{O + H}$$

$$= \frac{2\,250 \times 450}{370 + 450} \text{ nm}$$

$$= \frac{2\,250 \times 450}{820} \text{ nm}$$

$$= 1\,235 \text{ nm} \quad \ldots \text{(b)}$$

And time is 1 235 nm @ G/S 370 kt = 3 hr 20 min . . . (b)

Check

1 235 nm @ 450 kt = 2 hr 45 min
1 015 nm @ 370 kt = 2 hr 45 min.

PNR.

$$T = \frac{EH}{O + H}$$

$$= \frac{365 \times 450}{370 + 450} \text{ min}$$

$$= \frac{164\,250}{820} \text{ min}$$

$$T = 3 \text{ hr } 20 \text{ min} \quad \ldots \text{(c)}$$

And distance is 3 hr 20 min @ 370 kt = 1 233 nm . . . (c)

BUT we did not have to do this calculation for the PNR. As the PNR endurance is the flight time, the CP and the PNR should coincide!

Example 3

This is a PNR involving a return to base on 3 engines, and on several Tracks, practical enough, but there are pitfalls easily fallen into, if you are rushed for time.

An a/c is to fly from F to G via K and M; the data is as follows:

Stage	Wind component (kt)	Distance (nm)
F to K	+20	400
K to M	+15	630
M to G	+25	605

Mean TAS	500 kt
Mean TAS (3 engines)	435 kt
Mean fuel consumption (4 engines)	5 300 kg/hr
Mean fuel consumption (3 engines)	4 100 kg/hr
Fuel on board (inc. Reserve, 5500 kg, assume unused)	30 000 kg

Calculate the time and distance to the PNR from departure F, the return flight to F to be made on 3 engines.

Answer

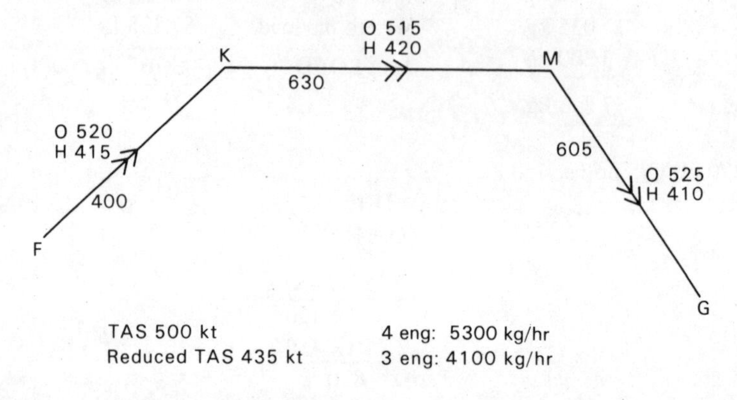

Fig. 13.3

Flight Out (TAS 500 kt)

	Wind comp	G/S	d	t	kg/hr	kg
F–K	+20	520	400	46	5 300	4 060
K–M	+15	515	630	$73\frac{1}{2}$	5 300	6 490
M–G	+25	525	605	69	5 300	6 100

Flight Back (TAS 435 kt)

	Wind comp	G/S	d	t	kg/hr	kg
K–F	–20	415	400	58	4 100	3 960
M–K	–15	420	630	90	4 100	6 150
G–M	–25	410	605	$88\frac{1}{2}$	4 100	6 050

Fuel analysis

	Out	Back	Out + Back	Total
F–K	4 060	3 960	8 020	8 020
K–M	6 490	6 150	12 640	20 660
M–G	6 100	6 050	12 150	32 810
	16 650	16 160		
		16 650	check	
		32 810		

Total fuel on board	30 000
PNR reserve	5 500
PNR fuel	24 500
Fuel K–M–K	20 660
Fuel M–PNR–M	3 840
Fuel M–G–M	12 150

So $\frac{3\,840}{12\,150} = \frac{d}{605} = \frac{t}{69}$

d = 191 t = 22

	d	t
F–M	1 030	$119\frac{1}{2}$
M–PNR	191	22
F–PNR	1 221	$141\frac{1}{2}$
	nm	min

∴ PNR is 400 + 630 + 191 = 1 221 nm from F.
and 46 + $73\frac{1}{2}$ + 22 = $141\frac{1}{2}$ min (2 hr $21\frac{1}{2}$ min) from F.

Example 4

On a trip from P to R via Q, an a/c is ordered in the event of turning back to proceed to its alternate Y via Q. TAS on 4 engines is 500 kt, on 3 engines is 420 kt.

Stage	Wind component (kt)	Distance (nm)
P to Q	−25	565
Q to R	−45	900
Q to Y	+30	240

(a) Give the time and distance from P to the three-engined CP between R and Y.

(b) FOB 38 000 kg, consumption 6 300 kg/hr, reserve (assume unused) 6 500 kg, and the whole flight is made on 4 engines, what is the distance from P to PNR to Y?

Answer

(a) CP

Full TAS (500 kt)

	wind comp	G/S	d	t
P–Q	−25	475	565	$71\frac{1}{2}$
Q–R	−45	455	900	118

Reduced TAS (420 kt)

	ON			BACK			
	w/c	G/S	t	w/c	G/S	t	d
Q–R	−45	375	144	+45	465	116	900
Q–Y				+30	450	32	240

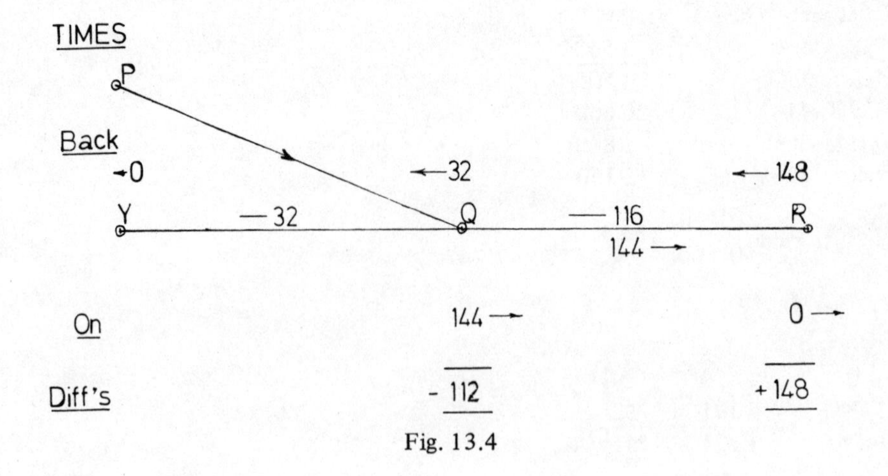

Fig. 13.4

From Fig. 13.4, CP is between Q and R.

$$\frac{112}{260} = \frac{d}{900} = \frac{t}{118}$$

$$d = 388 \quad t = 51$$

	d	t
P–Q	565	$71\frac{1}{2}$
Q–CP	388	51
P–CP	953	$122\frac{1}{2}$

∴ Time and distance from P to CP = 2hr $2\frac{1}{2}$min, 953 nm.

(b) PNR

Flight Out (TAS 500 kt)

	Wind comp	G/S	d	t	kg/hr	kg
P–Q	–25	475	565	$71\frac{1}{2}$	6 300	7 510
Q–R	–45	455	900	118	6 300	12 390

Flight Back (TAS 500 kt)

	Wind comp	G/S	d	t	kg/hr	kg
R–Q	+45	545	900	99	6 300	10 400
Q–Y	+30	530	240	27	6 300	2 840

Fuel analysis

	Out	Back	Out + Back	Total
P–Q	7510	–	10 350	10 350
Q–Y	–	2840		
Q–R	12 390	10 400	22 790	33 140
	19 900	13 240		
		19 900	check	
		33 140		

Total fuel on board	38 000
PNR reserve	6 500
PNR fuel	31 500
Fuel P–Q–Y	10 350
Fuel Q–PNR–Q	21 150
Fuel Q–R–Q	22 790

$$\frac{21\,150}{22\,790} = \frac{d}{900} = \frac{t}{118}$$

$$d = 835 \quad t = 109\tfrac{1}{2}$$

	d	t
P–Q	565	$71\frac{1}{2}$
Q–PNR	835	$109\frac{1}{2}$
P–PNR	1400	181

∴ Distance from P to PNR = 1 400 nm.

Example 5

There is a lot to be said for working out a full flight plan, completing it with a CP and PNR: after all, that is what would happen on the routes when you are forced to land at a field not manned by your own company's personnel.

Let us take one step-by-step with the trimmings left out:

A flight is to be made from A to D at cruise Mach 0·75: FOB 23 000 kg, reserve fuel (assume unused) 4 000 kg. Ignore descent.

Complete the flight plan and calculate

(a) time and distance for PNR to A

(b) time and distance for CP between A and D.

In our print-out of the flight plan, the data given has the column marked with an asterisk.

1. Evolve the temperature at FL from the temperature deviation column.
   Standard temperature at FL 330 is –51°C
   (33 000 ft @ 2°C per 1 000 ft = –66°C
   Standard at sea level = +15°C
   –51°C)

   ∴ A to B = –42
   B to C = –46
   C to D = –53

   Fill in the return temperature details.
2. Evolve TAS from Mach 0·75 from computer.
   Mach index in Airspeed window v ambient temperature and read speed of sound on outer scale against 1 on the inner.
   Do you agree consecutively 593, 587, 579, 583, 593, 600?

   Then $\frac{\text{TAS}}{\text{Sof S}}$ = Mach 0·75, or simply 593 on outer scale against 10 on inner,

FLIGHT PLAN

* STAGE		Temp.	*	*	* WIND		*						*		*	
From	To	°C	Flight Level	Temp. Dev. °C	Direction	Speed kt	Track ° (T)	Drift	Heading ° (T)	TAS kt	Wind comp. kt	G/S kt	Distance nm	Time min	Fuel flow kg/hr	Fuel required kg
A	B	−42	330	+ 9	230	60	290	7S	283	445	−35	410	430	63	4 000	4 200
B	C	−46	330	+ 5	250	70	303	8S	295	440	−49	391	365	56	3 700	3 450
C	D	−53	330	− 2	260	50	312	5S	307	434	−34	400	386	58	3 550	3 430
D	C	−50	310	− 3	270	40	132	3P	135	437	+31	468	386	$49\frac{1}{2}$	3 650	3 000
C	B	−42	310	+ 5	240	60	123	7P	130	445	+25	470	365	$46\frac{1}{2}$	3 750	2 900
B	A	−36	310	+11	220	50	110	6P	116	450	+15	465	430	$55\frac{1}{2}$	3 700	3 410

and read off 445 on outer against 0·75 on inner, and so on, and the TAS is as shown on the flight plan.

3. Complete the flight plan. We have 19 000 kg usable fuel, adequate for A to D and some suitable alternate.

The PNR will be:

Fuel analysis

	Out	Back	Out + Back	Total
A–B	4 200	3 410	7 610	
B–C	3 450	2 900	6 350	13 960
C–D	3 430	3 000	6 430	20 390
	11 080	9 310		
		11 080	check	
		20 390		

Total fuel on board	23 000
PNR reserve	4 000
PNR fuel	19 000
Fuel A–C–A	13 960
Fuel C–PNR–C	5 040
Fuel C–D–C	6 430

$$\frac{5\,040}{6\,430} = \frac{d}{386} = \frac{t}{58}$$

$$d = 303 \text{ nm} \qquad t = 45\tfrac{1}{2} \text{ min}$$

	d	t
A–C	795	119
C–PNR	303	$45\frac{1}{2}$
A–PNR	1098	$164\frac{1}{2}$ = 2 hr $44\frac{1}{2}$ min . . . (a)

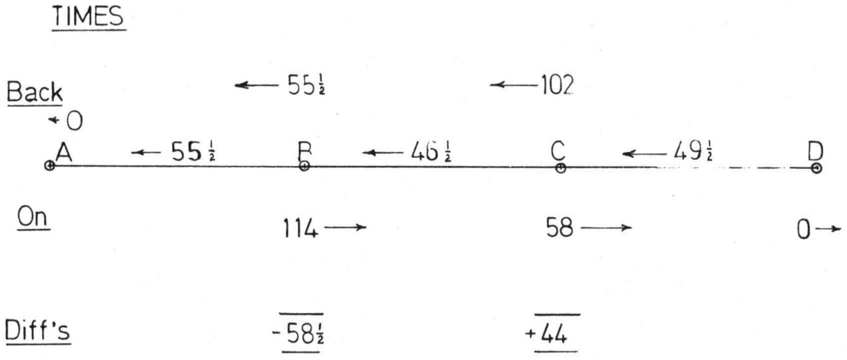

TIMES

Back ← 0 ← $55\frac{1}{2}$ ← 102

A ← $55\frac{1}{2}$ B ← $46\frac{1}{2}$ C ← $49\frac{1}{2}$ D

On 114 → 58 → 0 →

Diff's $-58\frac{1}{2}$ $+44$

Fig. 13.5

CP (see Fig. 13.5)

From Fig. 13.5, CP between B and C:

$$\frac{58\frac{1}{2}}{102\frac{1}{2}} = \frac{d}{365} = \frac{t}{56}$$

$$d = 208 \text{ nm} \qquad t = 32 \text{ min}$$

	d	t	
A–B	430	63	
B–CP	208	32	
A–CP	638	95	= 1 hr 35 min . . . (b)

Now a few worked examples which hurl you into the tables.

An aircraft is at FL 350 over aerodrome B: its weight is 123 000 kg, and it is cruising at Mach 0·86, mean headwind component 60 kt, temperature deviation –7°C. Fuel on board excluding reserves is 21 000 kg.

What is the range of the aircraft which will permit it to return to overhead B at the same FL? (Use Table 33C)

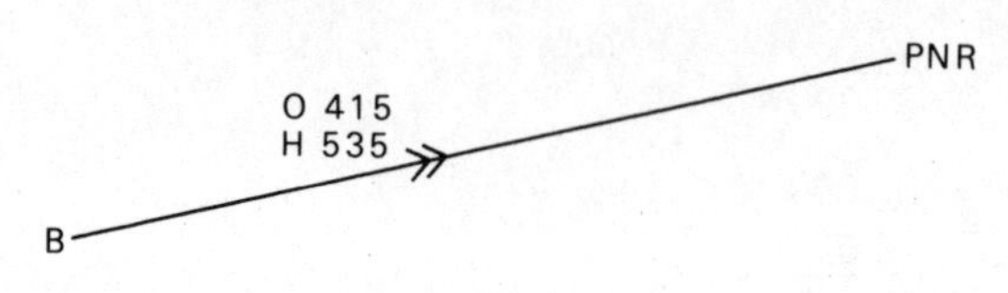

Fig. 13.6

Table 33C in the –10°C to –1°C temperature deviation set, with a lucky 123 000 kg all up weight gives

Fuel flow 6 700 for 1st hour ∴ 14 300 kg left
6 500 for 2nd hour ∴ 7 800 kg left
6 200 for 3rd hour ∴ 1 600 kg left

19 400 in 3 hours

and 1 600 kg at the 4th hour flow of 6 000 kg/hr will be used in 16 minutes. So Endurance is 3 hr 16 min.

TAS is 475K, so O and H can be inserted on diagram, and the formula entered.

$$T = \frac{EH}{O + H}$$

$$= \frac{196 \times 535}{415 + 535} \text{ min}$$

$$= \frac{196 \times 535}{950} \text{ min}$$

$$= 110 \text{ min or } 1 \text{ hr } 50 \text{ min}$$

and at 415 kt, the distance out is 760 nm.
A quick check will verify this.

We must try the other tables, so there follows a descent and hold job: for the holding, use Table 33D.

An aircraft is over its destination field (elevation 3 000 ft) at 30 000 ft, weight 88 000 kg, temperature deviation –3°C. The aircraft is instructed to hold at FL 140. Allowing fuel for circuit and landing from 1 000 ft of 1 000 kg, how long can it hold if fuel to be used before landing must not exceed 5 000 kg?

From the descent table (Table 33E), the descent from FL 300 to FL 40 (1 000 ft over the field) takes 12 min and uses 500 kg. Add to this the circuit and landing fuel of 1 000 kg, then we have (5 000 – 1 500) kg, 3 500 kg for holding.

In Table 33D, read off against Pressure Height and Temp. dev. the fuel flow 6 600 kg/hr. But watch the footnote: 1½% reduction in flow for every 5 000 kg in mean weight below 100 000 kg.

The hold will start at 87 700 kg AUW, since the descent from FL 300 to FL 140 uses 300 kg, a simple subtraction in Table 33E; and the hold will use 3 500 kg, so the mean weight will be (87 700 – 1 750) kg = 85 950 kg, call it 86 000 kg.

A decrease of 14 000 kg in weight = 4·2%, and 4·2% of 6 600 kg is 277 kg (we are keeping up the pedantic work). ∴ flow is 6 320 kg, rounding off the digits and 3 500 kg @ 6 320 kg/hr = 33 min, holding.

Keep going!
An aircraft diverts from 1 000 ft overhead its destination aerodrome (elevation 4 000 ft) to its alternate (elevation 2 700 ft). Weight is 75 000 kg, distance is 515 nm, mean wind component –45 kt.
The diversion is made at FL 320, temp. dev. –8°C.

(a) What fuel is required to overhead alternate (Table 33G)?
(b) Give mean TAS for climb from FL 200 to FL 320, temp. dev. –6°C (Table 33B).

(a) A spot of interpolation in Table 33G; against ground distances 510 nm and 520 nm, and in headwind component columns 40 and 50.

515 nm at –50 reads 7 485 kg, 85 min at FL 420.
515 nm at –40 reads 7 355 kg, 83 min at FL 420.
So at –45, use 7 420 kg, 84 min at FL 420.

Now for the corrections:

1. Start diversion at 5 000 ft — –300 kg
3. End diversion at 3 700 ft — Nil
3. Start diversion at 75 000 kg
   so subtract 1% of fuel for each 1 500 kg below 90 000 kg
   10% of 7 420 kg — –740 kg
4. Cruise at FL 320 for 84 min.
   If at FL 350, +240 kg for 84 min.
   If at FL 300, +1 020 kg for 84 min.
   ∴ 84 min at FL 320 = 1 020 – ($\frac{2}{5}$ x 780) = +700 kg

   –340 kg

   ∴ fuel required 7 080 kg ... (a)

(b) Table 33B.
Cross from 32 Top of climb height to the 20 000 ft curve, down to the reference line, then UP to –6°C temp. dev. Read 412 kt?

The following is quite a difficult problem; we have already done one of these, but not with an entrance into the tables, nor calling for some inspired guesswork. Use Tables 34C and 34G as appropriate.

An aircraft is to fly from PETALING to GLINKA, via KARVEL. Should an engine fail after overlying KARVEL, the return must be made to the alternate QUONTEK, via KARVEL.

The route details are as follows:

SECTOR	Distance (nm)	F/L	Wind Component (kt)	Temp. (°C)
PETALING to KARVEL	870	250	−20	−18
KARVEL to GLINKA	640	260	−30	−23
GLINKA to KARVEL	640	250	+30	−22
KARVEL to QUONTEK	700	260	−25	−23

Weight at start 280 000 kg, fuel available excluding reserves 61 000 kg: ignore climb and descent.

(a) If an engine fails after KARVEL, how far can it travel towards GLINKA before turning for QUONTEK as directed?

(b) How long after leaving PETALING will this point be reached?

Diagram first, and do not move without consulting it and the table of route details.

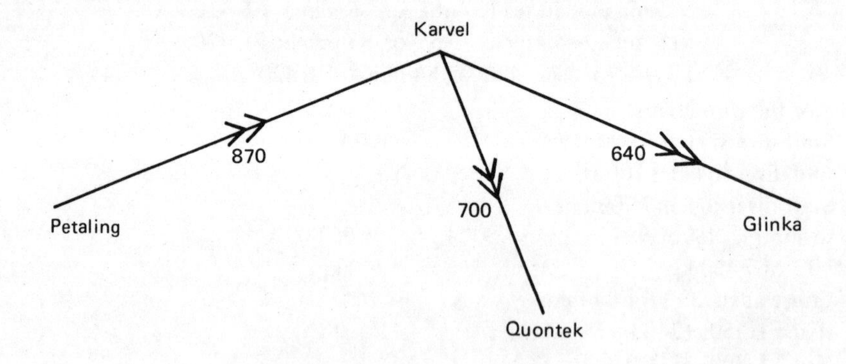

Fig. 13.7

Here is the working for the outward and return flights. The return flight was worked backwards (i.e. upwards from the bottom line of the flight plan) from the known landing weight at QUONTEK, i.e. take-off weight less the PNR fuel weight.

From	To	FL	Temp Dev	TAS	Wind comp	GS	d	t	Fuel Flow	Wt at Start (t)	Fuel Required
PET		250	+17	517	–20	497	870 { 497	60	13 900	280.0	13 900
	K	250	+17	517	–20	497	373	45	13 700	266.1	10 280
KAR		260	+14	510	–30	480	640 { 480	60	12 900	255.8	12 900
	G	260	+14	510	–30	480	160	20	12 700	242.9	4 230
GL IN		250	+13	444	+30	474	640 { 173	22	10 900		4 000
	K	250	+13	437	+30	467	467	60	10 500	246.3	10 500
KAR		260	+14	436	–25	411	700 { 297	$43\frac{1}{2}$	10 000	235.8	7 250
	Q	260	+14	428	–25	403	403	60	9 500	228.5	9 500
							TOW 280t – PNR fuel 61t =			219.0	

Fuel analysis

	Out	Back	Out + Back	Total
PET–KAR	24 180	–		
KAR–QUON	–	16 750	40 930	40 930
KAR–GLIN	17 130	14 500	31 630	72 560
	41 310	31 250		
		41 310		
		72 560 ← check → 72 560		

PNR fuel	61 000
Fuel P–K–Q	40 930
Fuel K–PNR–K	20 070
Fuel K–G–K	31 630

$$\frac{20\,070}{31\,630} = \frac{d}{640} = \frac{t}{80}$$

$$d = 406 \quad t = 51$$

K–PNR 406 nm . . . (a)

	t
P–K	105
K–PNR	51
P–PNR	156 min = 2 hr 36 min . . . (b)

APPENDICES

1: Glossary of Abbreviations

a/c	aircraft
A/D	aerodrome
ADF	automatic direction finding equipment
ADIZ	air defence identification zone
ADR	advisory route
agl	above ground level
A/H	alter heading
alt or Alt	altitude
amsl	above mean sea level
APS	aircraft prepared for service (weight)
ASI	airspeed indicator
ASR	Altimeter Setting Region
ATA	actual time of arrival
ATCC	Air Traffic Control Centre
ATD	actual time of departure
ATZ	Air Traffic Control zone
AUW	all-up weight
Brg	Bearing
BS	Broadcasting station
°C	degrees Celsius, hitherto called Centigrade
°(C)	degrees Compass
CA	conversion angle
CAA	Civil Aviation Authority
CAVOK	weather fine and clear
ch lat	change of latitude
ch long	change of longitude
CL	chart length
cm	centimetre(s)
C of G	centre of gravity
Comp	component
CP	critical point
C/S or c/s	call sign
CTR	control zone
Dev	deviation
DF	direction finding
DGI	directional gyro indicator
dist	distance
DME	distance measuring equipment

DR	dead reckoning
EAT	expected approach time
ETA	estimated time of arrival
ETD	estimated time of departure
ETW	empty tank weight
FIR	Flight Information Region
FIS	Flight Information Service
FL	flight level
FOB	fuel on board
ft	feet
ft/min	feet per minute
°(G)	degrees Grid
G/C	Great circle
GCA	Ground Controlled Approach
GD	Greenwich date
GMT	Greenwich Mean Time
Griv	grivation
G/S	ground speed
H24	operates 24 hours daily
Hdg	Heading
HF	High frequency
HJ	daylight hours
h m s	hours minutes seconds
hr(s)	hour(s)
ht	height
hwc	head wind component
Hz	Hertz (or) cycles per second
IAS	indicated airspeed
IFR	instrument flight rules
ILS	instrument landing system
IMC	instrument meteorological conditions
in	inch
ISA	International Standard atmosphere
Item A	the fuel from departure point to destination only
kg	kilogram(s)
kg/hr	kilograms per hour
kHz	kilohertz, or kilocycles per hour
km/hr	kilometres per hour
kt	knot(s)
Lat	Latitude
Ldg wt	landing weight
LD	Local Date; also landing distance
LF	Low frequency
LMT	Local Mean Time
Long	Longitude
LRC	Long Range Cruise
M	Mach

°(M)	degrees Magnetic
Mb/mb	millibar(s)
M–F	operates Mon. to Friday only
MF	Medium frequency
MHz	megahertz, or megacycles per second
min	minute(s)
M_{ind}	indicated Mach number
mm	millimetre(s)
MN	Mach number, Magnetic north
mph	statute miles per hour
msl	mean sea level
M/R	Moonrise
M/S	Moonset
NDB	non-directional radio beacon
NH	Northern hemisphere
NM/nm	nautical mile(s)
NP	North Pole
OM	outer marker
O/R	on request
PAR	Precision Approach Radar
PE	pressure error
P/L	position line
PNR	point of no return
posn	position
PP	pinpoint
PPO	prior permission only
Press Alt	pressure altitude
QDM, QDR, QNH, QTE	defined in the text
RAS	rectified airspeed
Rel	relative
R/L	Rhumb line
RLW	regulated landing weight
RMI	radio magnetic indicator
RTOW	regulated take-off weight
R/W	Runway
Rx	Receiver
SA	standard atmosphere
sg	specific gravity
SH	Southern hemisphere
S/H	set heading
sm	statute mile(s)
SP	South Pole
S/R	Sunrise
SRA	special rules area
SRZ	special rules zone
S/S	Sunset
SSR	Secondary Surveillance Radar

ST	Standard Time
Stn	Station
t	tonne
°(T)	degrees True
TAS	True airspeed
Temp	temperature
TMA	terminal control area
TMG	Track made good
TN	True North
T/O	take-off
TOC	top of climb
TOD	Top of descent, also take-off distance
TOW	take-off weight
Tr	Track
TVOR	terminal VHF omni-directional range
twc	tail wind component
Tx	transmitter
UKIAP	United Kingdom Aeronautical Information Publication, known as the UK Air Pilot
UHF	Ultra high frequency
u/s	unserviceable
Var	variation
VDF	VHF direction finding
VFR	visual flight rules
VHF	very high frequency
vis	visibility
VMC	visual meteorological conditions
V_{NE}	never exceed speed
V_{NO}	normal speed
VOR	VHF omni-directional range
W/D	wind direction
Wind comp	wind component
W/E	wind effect
W/S	wind speed
wt	weight
W/V	wind velocity
ZFW	zero fuel weight

2: Conversion Factors

Imp gal	to	litres	multiply by 4·546
litres		Imp gal	0·22
Imp gal		US gal	1·205
US gal		Imp gal	0·83
gal		cubic ft	0·161
cubic ft		gal	6·25
lb/sq in		kg/cm^2	0·07
lb		kg	0·454
kg		lb	2·205
ft		metres	0·3048
metres		ft	3·2808
sm		nm	0·8684
nm		sm	1·1515
sm		km	1·609
km		sm	0·621
nm		km	1·852
km		nm	0·54
in		mb	33·86
mb		in	0·0295
°C		°F	use formula $(°C \times \frac{9}{5}) + 32$
°F		°C	use formula $(°F - 32) \times \frac{5}{9}$

3: Navigation Equipment, Charts, etc.

Plotting gear: can be bought at a number of shops providing for draughtsmen, but is best obtained from those specialising in pilots' requirements;
Airtour International, at Elstree Aerodrome, Herts WD6 3AW has a large stock; they have other branches.

Dividers: buy the compass-divider type (those vast contraptions that look like instruments for getting the tops off bottles of pickles are strictly for the yachties).

Protractor: 5-inch square (eg. Airtour PP2)

Rule: a 30 cm clear plastic ruler is a good investment, with inches, tenths, and centimetres, millimetres.
Scale ruler: 20″ showing nm for 1:1 000 000 scale (eg. Airtour NM4)

Computer: there are numerous types, avoid movable wind-arms, make sure it goes up to high speeds and has all the refinements like sg, Mach, etc. on the circular slide rule (e.g. Airtour CRP 5).

Electronic calculators: simple scientific as recommended for GCE 'O' level. It is worth considering a model with solar cells so removing the dependence on batteries.

Maps and charts: obtainable from CAA Chart Room or accredited agents (see green AIC) including
Edward Stanford Ltd, 12–14 Long Acre, London, WC2E 9LP
Airtour International, as above.
Instructional Plotting Chart UK (Lambert's) 1:1 000 000 and
Instructional Plotting Chart Europe (Lambert's) 1:1 000 000 also available from CAA Chart Room, CAA House, 45–59 Kingsway, London WC2B 6TE.

Aeronautical Information Circulars are obtainable from Aeronautical Information Service, Tolcarne Drive, Pinner, Middlesex, HA5 2DU on payment of annual postage. The circular on aviation charts is a handy reference, and it has a list of chart symbols.

Data Sheets (currently numbers 33 and 34) for Flight Planning, CAP 505 Objective Testing for Professional Licences, CAP 519 Objective Testing for Pilots' Instrument Rating are all from CAA, Greville House, 37 Gratton Road, Cheltenham.

Aerad charts: the useful one for this book is any fairly recent EUR 1/2. Available from British Airways AERAD Customer Services, AERAD House, Bldg 254/490. Heathrow Airport (London), Hounslow, Middlesex TW6 2JA or Airtour.

Jeppesen Enroute Charts: The principal symbols which appear on Jeppesen enroute charts which differ from those on Aerad comprise:

1. Indication of FLs. If FLs disagree with the semicircular rules, on Jeppesen the routes are annotated O> or E> depending upon whether ODDs or EVENs are in use respectively.
2. Total distances between radio facilities. In addition to distances between reporting points and/or sector points, Jeppesen include values in hexagons to show the total distance between successive radio facilities.

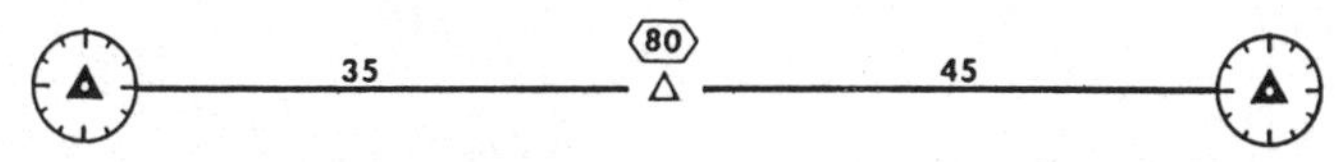

3. DME ranges. The letter D indicates DME range in nm, either as

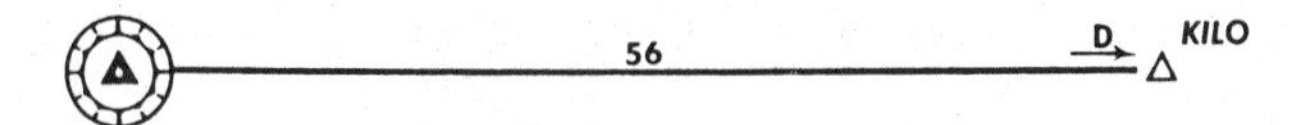

indicating that KILO would be determined by radial and DME range 56 nm from the station, or as

KILO D95 274°
XYZ
114.5

where KILO is on radial 274 at a range of 95 nm from XYZ DME/VOR, frequency 114.5.

4. Direct tracts. A capital D in a blue square is used.
5. Danger/Restricted/Prohibited Areas. Identity includes ICAO 2-letter locator, and, unless space prohibits, the activity is specified in the area on the chart. SFC means surface. In •EG(D)136, the • indicates always active, as scheduled.
6. Minimum off-route altitudes. (MORA). Although they may be shown as grid values (1° lat/1° long), MORA are given in feet along routes followed by a suffix a. e.g. 3500a. The clearance within 10 nm of the route centre-line given by MORA from terrain and known obstacles is

   1000ft for MORA up to 7000a and
   2000ft for MORA greater than 7000a.

7. Radio frequency sector boundary. Shown on Jeppesen by an outline of 'telephone hand-sets' in green: . Within the boundary, the common authority callsign and frequency are given, an asterisk (*) indicating not H24 service and (R) indicating Radar.

Answers to multi-choice Test Questions

Chapter 3
Q1 (c), Q2 (b), .Q3 (b), Q4 (a), Q5 (d), Q6 (c), Q7 (c), Q8 (a), Q9 (d), Q10 (d), Q11 (a), Q12 (d), Q13 (d), Q14 (a), Q15 (d), Q16 (c), Q17 (d), Q18 (c), Q19 (a), Q20 (b).

Chapter 4
Q1 (a), Q2 (b), Q3 (c), Q4 (c), Q5 (b), Q6 (c), Q7 (b), Q8 (c), Q9 (c).

Chapter 11
Q1 (b), Q2 (d), Q3 (c), Q4 (a), Q5 (b), Q6 (d).

Chapter 12
Q1 (a), A2 (d), Q3 (a), Q4 (d), Q5 (b), Q6 (c), Q7 (d), Q8 (b).

Index